Florida's Space Coast

THE FLORIDA HISTORY AND CULTURE SERIES

UNIVERSITY PRESS OF FLORIDA

Florida A&M University, Tallahassee
Florida Atlantic University, Boca Raton
Florida Gulf Coast University, Ft. Myers
Florida International University, Miami
Florida State University, Tallahassee
New College of Florida, Sarasota
University of Central Florida, Orlando
University of Florida, Gainesville
University of North Florida, Jacksonville
University of South Florida, Tampa
University of West Florida, Pensacola

THE IMPACT OF NASA ON THE SUNSHINE STATE

WILLIAM BARNABY FAHERTY, S.J.

Foreword by Raymond Arsenault and Gary R. Mormino

University Press of Florida
Gainesville
Tallahassee
Tampa
Boca Raton
Pensacola
Orlando
Miami
Jacksonville
Ft. Myers
Sarasota

FLORIDA'S
SPACE
COAST

First cloth printing, 2002
First paperback printing, 2024

29 28 27 26 25 24   6 5 4 3 2 1

Library of Congress Cataloging-in-Publication Data
Faherty, William Barnaby, 1914-
Florida's space coast: the impact of NASA on the Sunshine State / William Barnaby Faherty;
foreword by Gary R. Mormino and Raymond Arsenault.
p. cm. — (The Florida history and culture series)
Includes bibliographical references and index.
ISBN 978-0-8130-2563-6 (cloth)   ISBN 978-0-8130-8055-0 (pbk)
1. John F. Kennedy Space Center—History. 2. Brevard County (Fla.)—Social life and
customs—20th century. 3. Florida—History—20th century. 4. United States. National
Aeronautics and Space Administration—History. I. Title. II. Series.
TL4027.F52 J6377 2003
629.47'8'0975927—dc21   2002028931

The University Press of Florida is the scholarly publishing agency for the State University
System of Florida, comprising Florida A&M University, Florida Atlantic University,
Florida Gulf Coast University, Florida International University, Florida State University,
New College of Florida, University of Central Florida, University of Florida, University of
North Florida, University of South Florida, and University of West Florida.

University Press of Florida
2046 NE Waldo Road
Suite 2100
Gainesville, FL 32609
http://upress.ufl.edu

To my many Florida friends

Geography chose the East Coast of Florida
for the spaceport with the same certainty
that it chose the Isthmus of Panama
for the Great Canal.

FOREWORD

Florida's Space Coast: The Impact of NASA on the Sunshine State is the twenty-first volume in a series devoted to the study of Florida history and culture. During the past half century, the burgeoning population and increasing national and international visibility of Florida have sparked a great deal of popular interest in the state's past, present, and future. As the favorite destination of countless tourists and as the new home for millions of retirees and other migrants, modern Florida has become a demographic, political, and cultural bellwether. Unfortunately, the quantity and quality of the literature on Florida's distinctive heritage and character have not kept pace with the Sunshine State's enhanced status. In an effort to remedy this situation—to provide an accessible and attractive format for the publication of Florida-related books—the University Press of Florida has established the Florida History and Culture Series.

As coeditors of the series, we are committed to the creation of an eclectic but carefully crafted set of books that will provide the field of Florida studies with a new focus and that will encourage Florida researchers and writers to consider the broader implications and context of their work. The series will continue to include standard academic monographs, works of synthesis, memoirs, and anthologies. And while the series will feature books of historical interest, we encourage the submission of manuscripts on Florida's environment, politics, literature, and popular and material culture for inclusion in the series. We want each book to retain a distinct personality and voice, but at the same time we hope to foster a sense of community and collaboration among Florida scholars.

The publication of *Florida's Space Coast: The Impact of NASA on the Sunshine State* fills an important chapter in the history of the Sunshine State. The author, William Barnaby Faherty, S.J., brings to this book the talents of a distinguished historian, but also the perspective of a scholar who has observed and written about Florida's space program since the 1960s.

The take-off (literally and figuratively) of Cape Canaveral and Brevard County represents one of the most dramatic stories of modern Florida and Cold War America. On the eve of World War II, Cape Canaveral and Brevard County were two of America's most isolated and undeveloped locales. In 1940, Brevard County numbered only 16,142 residents. A few cities and towns—Melbourne, Cocoa, and Titusville—were scattered across the county's 1,000 square miles.

Soon, the world recognized dateline Cape Canaveral, known previously to scholars of colonial Florida as the site named by Ponce de León on his 1513 voyage of exploration. The Cape's transformation was swift and dramatic. Liberal applications of DDT and federal funds helped tame the voracious mosquitoes and turn swamp and grove into a site for rocket testing and guided missiles. On July 20, 1950, engineers and technicians cheered the first successful missile launch at Cape Canaveral.

Florida's Space Coast chronicles the evolution of Cape Canaveral from a primitive facility comprising a handful of scientists into a megacomplex. Faherty's book encompasses the arrival of the National Aeronautics and Space Administration, the Apollo and Mercury programs, and the renaming of Cape Canaveral (to Cape Kennedy) and the restoration. In addition, Faherty helps readers understand the remarkable transformation of Brevard County into a metropolitan and economic power. Throughout the 1950s, 1960s, and 1970s, Brevard County was one of America's fastest growing counties, the home to engineers, businesses, and technological firms. Professor Faherty is to be commended for this valuable book.

Raymond Arsenault and Gary R. Mormino
Series Editors

"The National Aeronautical and Space Administration has been one of Florida's most welcome residents," Governor Reubin O'Donovan Askew proclaimed on September 28, 1973, "providing jobs for thousands and spurring the development of a massive scientific and technical business community in our state." Fifteen years before, on October 1, 1958, the Congress of the United States, with the approval of President Dwight D. Eisenhower, had set up NASA to explore space for the benefit of all mankind. Three years after that, in May 1961, the Space Agency accepted the challenge of President John F. Kennedy to send a man to the moon and bring him back safely within the decade. By 1973, it had done more than that. Before the end of the century, it had done even more.

With the cooperation of the armed services, the nation's universities, and countless business firms, NASA brought 25,000 skilled men and women and their families to the area and sent unmanned launches that put weather satellites in orbit around the earth to provide more accurate weather reporting and communications satellites to bring the continents closer by phone and television. It sent probes that landed on the moon. It sent twelve men to the moon and brought them back safely. It explored other celestial bodies and pushed out into the vast reaches of the universe. It launched one hundred reusable spacecraft, the Shuttle, and it is, at the time of this writing, equipping a station in space, a launching site to unfathomable possibilities. The space program changed the face of Brevard County in east central Florida and readied the entire state for life in a new century and a new millennium.

I come to my subject—NASA's impact on Florida's development—a veteran historian and Apollo mission chronicler. Shortly before Governor Askew's speech in 1973, the History Department at the University of Florida recruited me as the senior member of a two-man team to write the history of the Apollo moon launches. This agreement between the department's Dr. David Bushnell and the National Aeronautics and Space Agency (NASA) had one major requirement. The senior team member had to be a full professor with at least one published book to his credit. By that time a full professor at Saint Louis University, I'd published three books: one on women's rights, called *The Destiny of Modern Woman* (1952) that anthropologist Margaret Mead reviewed favorably; the history of my *alma mater, Saint Louis University and Community, 1818–1968* (1968); and an historical novel, *A Wall for San Sebastian* (1962) that Metro-Goldwyn-Mayer adapted for the Anthony Quinn/Charles Bronson film *Guns of San Sebastian* (1969). I sought a sabbatical from Saint Louis University and moved to Titusville, Florida, in January 1972.

As the other member of the team, Dr. Bushnell enlisted Dr. Charles Benson, a member of Phi Beta Kappa, who had just completed his doctorate in history at the University of Florida. We began our work at Kennedy Space Center near Cape Canaveral in 1972. The challenge exhilarated us Apollo historians. We were giving our time and skills to the country in gratitude for all it had done for us—Dr. Benson at the outset of his career as an historian, and I in mid-passage. We were caught up in an atmosphere of transcendent achievement. We were to tell of a feat that ranked with the journey of the ancient Phoenician navigators through the Gates of Hercules into the North Atlantic, the voyage of Columbus to the New World, and the return of Magellan's ships from the first circling of the globe.

With the assistance of countless people at the Kennedy Space Center and the support of Professors Bushnell, John Mahon, and Michael Gannon in Gainesville, we finished our work at Kennedy Space Center in two years. NASA published the results of our research and writing in 1976 under the title *Moonport: A History of Apollo Launch Facilities and Operations*. The book received many reviews over the years, all favorable.

Dr. Benson remained at Kennedy Space Center to work on the history of Skylab, while I resumed my teaching at Saint Louis University.

Since then I have written twenty-two other histories of local institutions, such as the Missouri Botanical Garden. This book won the "Best Book of the Year Award" from the Missouri Writers' Guild in 1988. I taught night classes at Parks College of Aeronautical Technology and wrote the story of the school entitled *Parks College: Legacy of an Aviation Pioneer.* NASA recruited many graduates of that school, among them Walter Cooney, systems engineer at Kennedy Space Center, Ray Cerrato, chief of research and development, and Eugene Kranz, director of flight operations at Houston, who brought the Apollo 13 astronauts safely home.

Since the publication of *Moonport*, my interest in Florida's space coast has remained keen. I have revisited the area every year, met with many veterans of the Apollo days, and kept in touch with books on space exploration.

The movie *Apollo 13* galvanized the attention of the entire nation once again on the moon launches and pointed out the great work of Eugene Kranz at the controls in Houston. At this juncture, Meredith Morris-Babb, editor-in-chief of the University Press of Florida, published a second edition of *Moonport: A History of the Apollo Launch Facilities and Operations*, in two paperback volumes, *Gateway to the Moon* and *Moon Launch*, in 2001. The reception of these books called for more on the space story. Ms. Morris-Babb invited me to write a book on the influence of NASA on the development of the state as part of the Florida History and Culture Series.

In the meantime, many books had come out on Mercury and Gemini flights, on NASA and its predecessor, the National Advisory Committee on Aeronautics (NACA), on the development of the Saturn and the command module, on the hopes, plans, and proposals of the dreamers, planners, and administrators, on the experiences of the astronauts, on the Skylab, on the Space Shuttle, and on the many unmanned launches. Readers heard the story through the eyes of space pioneer Wernher von Braun and astronaut Michael Collins, but not from the viewpoint of the workers who put the vehicle in space. Editor-in-chief Meredith Morris-Babb approved my approach to tell the story from the viewpoint of Fred Renaud who drove the crawler-transporter and Clarence Chauvin who conducted tests for Apollo, from the experiences of their wives Pat Renaud and Mona Chauvin, of non-space personnel such as home-builder Donald Boland and Charles Wintgersahn, who looked to good living in Brevard County, and their wives Chris Boland and Rose Wintgersahn, and from the vantage of Irene McConarty and Sister Margaret Mary Kenny, who taught their children.

No book covered all the manned launches from Mercury to the Shuttle, and unmanned launches from the hit-and-miss early attempts, through communications and weather satellites to exploring the planets. Even more significantly, no one told the story of the influence of the entire space program on development of the state. This I set out to do in the spring of 2001.

Acknowledgments

My thanks go first to the editor-in-chief of the University Press of Florida, Meredith Morris-Babb, for inviting me to write this book for the Florida History and Culture Series; to historian Dr. Michael Gannon of the University of Florida for suggesting my name to Ms. Morris-Babb and for evaluating the completed manuscript; to Dr. Roger P. Launius for urging that I enlarge the bibliography and for making available "An Annotated Bibliography of the Apollo Program," Washington, NASA, 1994; to Dr. Charles D. Benson; to Raul Ernesto Reyes, NASA test conductor, a veteran of the Apollo moon launches, and Dan Venverloh, engineer with McDonnell Douglas, who went over several drafts of the manuscript from their vantages; and the many other Floridians who shared their experiences with me: Jean Charron, Mike Menghini, James Voor, Ida Reyes, Wallace Dal Santo, Vern Jansen, Art Gruenenfelder, Donald Boland, Rev. Liam Tobin, Fred Renaud, Drs. Robert Gray and Thomas O'Malley, Lt. Col. Lee Verbeck, William Horner, Ralph Hall, William Murphy, and George White, whose articles in the *Star Advocate* alerted Brevardians to this project.

I also wish to thank the archivists and librarians who supported my research: Janice Taylor and Cassy Adams of the Titusville Public Library; Elaine Liston, Kay Grinter, Barbara Green, and others at the Kennedy Space Center Archives; Cynthia Hudson and Dr. Miriam Joseph of the Pius XII Library in St. Louis, Missouri; John Waide of the Saint Louis University Archives; historian Lori Walters of Deltona, Florida;

Dr. Kathy O'Donnell of the U.S. Air Force Academy; and William L. Mugan, S.J., director of the Midwest Jesuit Archives. Special thanks go to Mary Struckel, who typed and edited drafts and the final copy of this manuscript; to Nancy Merz, associate archivist, who appraised the first and final copies; and to the priests and people of St. Theresa's Parish in Titusville, whose friendship helped to make my two-year stay in Florida so enjoyable.

All of the photographs in this book are courtesy of NASA and the NASA history office and can be found at http://grin.hq.NASA.gov.

()

TESTING THE TERRITORY

White Sands, but No Seashore

Slightly populated Brevard County, while only fifteen to twenty miles wide, stretched for seventy miles along Florida's east coast, midway between Jacksonville and Miami. Its main feature, an elbow of land called Cape Canaveral, jutted out into the Atlantic. The Cape covered forty square miles. Early Spanish sailors, who came up the coast and then moved out into the North Atlantic, noted the abundance of cane reeds, and gave the area its name, Canaveral, the Spanish name for a thicket of canes. It was the second-oldest place name in the United States, next to that of Florida itself.

While most vacationers and retirees went through the area to Palm Beach or Miami, North Brevard residents turned from hunting and fishing to nourishing citrus groves. Industry barely had a toehold. Only one firm, Harris Electronics in Melbourne, employed over twenty workers. With the heaviest population near Melbourne in the south and the county seat Titusville in the north, Brevard County had no central focus. But it did have room. It covered almost as many square miles as the State of Rhode Island.

The people of Titusville looked directly east over the Indian River, Merritt Island, and Banana River to the Cape. A larger community, Daytona Beach, hugged the Atlantic fifty miles to the north. Orlando lay thirty miles beyond St. John's River, the county's western boundary. Organized in 1855, only ten years after Florida entered the Union, the county bore the name of Judge Theodore Washington Brevard, Comptroller of the Sunshine State.

In October 1946, a committee of the Joint Chiefs of Staff decided that Cape Canaveral in Brevard County was the place for the military services to test guided missiles. Several factors dictated this choice. The flights went over the sea, not over inhabited territory. They could go southeast more than 5,000 miles into the South Atlantic. Many Caribbean islands offered sites for electronic equipment to monitor the flights. The Banana River Naval Air Station served as a support base, and the launch area was accessible to water transportation.

This decision set in motion a series of events that over a few years brought 25,000 engineers to the area—more individuals than had previously lived in the county. When the first space-workers arrived in 1947, they found no shopping centers, schools, churches, libraries, or theaters, and few residences. They had to drive on a narrow road to Orlando to purchase the things they needed. And yet, 100,000 people, the spaceport workers and their families, were to come in a short time. Besides servicemen and personnel of the National Aeronautics and Space Administration (NASA), set up in the meantime, numerous companies had their part in this tremendous enterprise—Boeing, McDonnell, Douglas, Pan American, TWA, IBM, RCA, Convair, Grumman, Lockheed, Rockwell, General Dynamics, General Electric, Bendix, North American Aviation, and Martin, among others, and their suppliers and subsidiaries. In a short time, the area was to be an internationally known center for space exploration.

The story of the spaceport and its influence on the development of the State of Florida begins in one aspect far away on the dry uplands, near Roswell, New Mexico. Back in the 1920s, an American scientist, Robert Goddard, after facing opposition in his native New England, had moved there to experiment with rockets. Over the succeeding years, Americans showed interest in other areas of thrust and power. Unfortunately, none of his countrymen carried on Goddard's work.

In the Weimar Republic of Germany, however, a group of young men were moving into this field in the late 1920s. The Versailles Treaty had greatly restricted the development of German armament. It limited the number of airplanes and forbade the development of heavy artillery, but ignored rocketry. As a result, some of the brighter young men in the German scientific community, led by Dr. Wernher von Braun, experimented with rockets. Von Braun was a protégé of Herman Oberth, a German scientist born in the Transylvanian Province of the old Austro-Hungarian Empire, who, in 1923, published a classic study, *Die Rakete zu den Planetenräumen (Rockets in Planetary Space)*. During the 1930s and on into World War II, these young men with von Braun tried new theories and ultimately developed several weapons, called V-1 and V-2, that caused havoc during and shortly after the invasion of Normandy in 1944. In order to make sure that these men and their rockets did not end up in the hands of the Soviets at the fall of Germany, the American military dispatched special forces into Germany to bring out a number of the V-2 rockets and sent General John B. Medaris to make contact with the von Braun team that had developed these rockets at Peenemünde on the Baltic Coast.

Tall, erect, and square-shouldered, Dr. Wernher von Braun had the vision of novelist Jules Verne, the diplomatic persuasiveness of a contemporary, U.S. Senator Lyndon Baines Johnson from Texas, and the engineering and administrative skills of James Eads, who built the ironclads that opened the Mississippi River during the Civil War and erected the great bridge that spanned its waters shortly after the conflict. Like these three men, von Braun put his full attention and unusual skills to the new task.

It is strange, in looking back, to find that the men who had almost brought disaster to the Allied forces in the invasion of Western Europe should now be given opportunities in America to develop more powerful rockets. One must not forget, however, that in international affairs, the *next* war is always the important one. Japan, for instance, was an ally of England and America in World War I, an enemy in World War II. Twenty years later, navy officials invited the Japanese naval officer who planned the devastating attack on Pearl Harbor to address fledgling American naval officers. Military machines have turned on their allies shortly after a victory or recognized, as the United States did, that our former ally Russia, not Germany, threatened stability in Western Europe. And so, the American Army brought these Germans to America and gave them the opportunities to develop more sophisticated rockets. Florida was to have a significant part in that development.

To bring German rocket specialists, men who a year before had been using their genius to try to stop the Allied efforts for victory in Europe, into an area of the United States where men had fought and died in that fight against Nazi power, seemed a hazardous undertaking. One could imagine occurrences of violence against scientists of Germanic background. The hiring agencies were aware of this. José González of El Paso, Texas, who worked with the von Braun team from the earliest days, explained the procedure to avoid trouble. Those who hired American veterans to work with the German scientists, General Electric among them, were cautious. They asked where a candidate had served in the war, if he had any brothers or cousins who served in the European theater, and whether they were killed there. Anyone who failed to pass this security test went into areas of work not involving citizens of the former enemy. In this way, things moved reasonably well. There was little evidence of friction between the Americans and the new-coming German rocketeers.

General Medaris chose the area of White Sands, between Alamogordo and Las Cruces, New Mexico, near to the place where Robert Goddard had worked. In this less populated area, noise and strange flying objects were of less concern than they might have been in other areas of the country, such as Dr. Goddard's native New England.

A resident of Cocoa Beach, Florida, Frank Ochoa-Gonzalez, who grew up in Las Cruces, recalls hearing his grandfather talk of the coming of the missiles and the missile-men. In mid-August 1945, 300 railroad freight cars brought V-2 components to the Southwest. Frank's grandfather worked on the 46–foot V-2 rocket during its frequent tests in 1946 and 1947.[1]

On May 29, 1947, a rocket chose its own flight pattern and landed in a cemetery near Juarez, Mexico. The president of that country had already vetoed flights from California that threatened Mexican territory. Would the *gringos* ever learn? "German scientists," a comedian stated, "we're flying German birds onto Mexican territory and we paid. Montezuma was taking revenge in an unusual way." An eight-year-old boy in El Paso, Raul Ernesto Reyes, outside his home in a bus, heard the thud of that rocket across the Rio Grande. He had already seen the missiles trucked through El Paso, and he sensed the meaning of the noise. Little did he dream that twenty years later he might conduct tests of rockets destined to send men to the moon.

A Primitive Spaceport

The von Braun team, officially the Army Ballistic Missile Agency, had to find two new homes, one to develop missiles, the other to test them in flight. The town of Huntsville, on the Tennessee River in the north central section of Alabama, had housed the Redstone Arsenal since 1941, the first year of World War II. The place had many advantages for developing missiles. The Army had used it to produce and test weapons for chemical warfare. By 1947, it went on sale as surplus property. Alabama Senator John Sparkman, local Congressman Bob Jones, and businessmen of the area looked for a new tenant. They pointed to the advantages of the area: (1) the existing arsenal facilities with an area of 425.5 square miles; (2) abundant low-cost electric power from the TVA; (3) adequate water supply from the Tennessee River; and (4) open space for future development. Further, the Tennessee River offered a waterway to transport materials and finished rockets via the Ohio and Mississippi Rivers, and ultimately the Gulf of Mexico, and around the Keys to reach Cape Canaveral. The missile team moved, then, to the Redstone Arsenal at Huntsville, Alabama, in 1950, to continue its rocket development.

Among the members of the team going to Alabama, von Braun chose five New Mexicans of Hispanic background: Rodolfo Barraza, Antonio Beltran, Ramon Samaniego, Carlos Marshall, and the already-mentioned José González. In recruiting these skilled men of an often-spurned minority, von Braun anticipated fair employment practices that came to be implemented by many government agencies and private industries in the following years. Not too long after settling in northern Alabama, the von Braun team began to invite math and science teachers from the high schools of the region to visit Huntsville. Those who took advantage of the offer returned to their communities and talked of interplanetary travel and landing men on the moon to mostly unbelieving audiences. They might just as well have said that von Braun and his missile-men were planning to level Appalachia and deposit four feet of black loam on those once rocky areas to enrich the countryside with alfalfa fields.

In the meantime, on September 1, 1948, the newly independent Air Force had taken over the Banana River Naval Air Station. The Navy had opened the station shortly after the attack on Pearl Harbor to counteract enemy submarines along the Florida coast. The Air Force planned to use it as the headquarters for the joint long-range proving ground, and set up Patrick Air Force Base there. It lay twenty miles south of Cape Canaveral, an ideal place to launch missiles. On May 11, 1949, President Harry S. Truman signed a bill that called for the development of a missile range, to be called the Long Range Proving Ground. Slightly more than a year later, the Department of Defense assigned the responsibility for the range to the Air Force. The Coast Guard opened to missile use its installation on Cape Canaveral that covered almost a square mile.

This scenic but previously unsettled place, Cape Canaveral, with beautiful beaches, excellent fishing areas, a lighthouse, and scattered private residences, became the Air Force headquarters. The area had a few unpaved roads or trails, a dock used by shrimpers, and wildlife, welcome and unwelcome, including deer, alligators, rattlesnakes, and many millions of the pests that gave their name to the Mosquito Lagoon to the north. Back in the 1920s, realtors had great plans. The Depression thwarted them. One little village, Avon-by-the-Sea, stood near Winslow Beach.

Design engineer James E. Finn, one of the first men to come down to the Cape Canaveral area, recalled the original launch pad. It was a twenty-yard wide layer of concrete, poured on top of sandy soil, less than a mile north of the lighthouse on Cape Canaveral. When dozens of jeeps and delivery trucks sank to their axles on the sandy paths that filled in for roads, a layer of gravel was laid over the sand. Steel scaffolding purchased from painters surrounded the missile to form the first gantry or service support tower. Plywood platforms stood at various levels of the scaffolding. If more than ten missile-men climbed the piping at the same time, the whole rickety framework seemed ready to fall down. Finn belonged to the crew that stacked sandbags around an old shack—a one-time dressing room for swimmers—and turned it into a launch control blockhouse. It stood less than 100 yards from the pad. A row of trailers contained additional facilities to coordinate countdown information and reports from tracking sites. It was hot and humid, and mosquitoes seemed intent on driving all human beings back whence they came.

Missile-men prepared to inaugurate this primitive spaceport on July 19, 1950 with a two-stage missile, dubbed Bumper 7. It consisted of a modified V-2 first stage, and a WAC Corporal second stage, developed at the Jet Propulsion Laboratory of the California Institute of Technology. The launch crew, along with representatives of the Army, General Electric Corporation, and the California Institute of Technology, waited on the beach with 100 newsmen. The newsmen lacked tents, chairs, writing desks, and other facilities. The heat had sapped their enthusiasm. Bumper 7 sputtered and fizzled. Salt air had corroded some of its elements. A few newsmen left, but most stayed. Five days later, the launch crew tried again with a sister missile. Thunder roared across the Cape. Bumper 8 rose steadily into the air. At 15,500 meters, the WAC Corporal second stage ignited and accelerated to 2,703 miles per hour before dropping into the sea. America had taken its first step in the space age.

The Cape area had its growing pains, as all branches of the military sought preference. In response, on June 30, 1951, the Defense Department put the Air Force in sole charge of "the Cape Canaveral Missile Test Annex." All the while, American officials were negotiating with Great Britain to develop a tracking system that would go to Ascension Island in the South Atlantic, more than 5,000 miles southeast of Cape Canaveral.

When troubles arose over a division of operation between Patrick Air Force Base and the launch site twenty miles away, Pan American World Airways, an old hand at operating air bases around the world, convinced the Air Force that it could reduce the costs of running the range. The Air Force awarded Pan Am a contract for day-to-day operations. Soon Pan Am moved beyond the setting up of cafeterias to the provision of security on the pads. The Radio Corporation of America, in turn, received a subcontract for technical aspects of range operations.

The first Air Force winged missile flown from Florida, the Matador, built by the Martin Company with a range of 650 miles, went up on June 20, 1951. In succeeding years, 153 other Matadors passed the flight tests. Another pilotless bomber, the Snark, made ninety-seven flights totaling 5,000 miles down the range after August 1952. Martin later developed the Mace, an improved and longer-ranged version of the Matador. The August 1953 launch of the Army's Redstone signaled the end of winged missiles.

The Missile Firing Laboratory at Huntsville, under the direction of Dr. Kurt Debus, a professor at Darmstadt in Germany before becoming a member of the von Braun team at Peenemünde during the war, regularly journeyed to the Cape for launches. Frank Childers, an expert in quality control, recalls that thirty-seven men came in a caravan from Huntsville in 1953 and stayed. Their families came down the following year. A short time later, the Missile Firing Laboratory took up permanent residence in the vicinity of the Cape.

In the late fifties, the U.S. began to launch its intercontinental ballistic missiles from a Cape area. The Thor, a sixy-five-foot-tall 110,000-pounder with a range of 1,700 miles, built by Douglas Aircraft Company under contract with the Air Force, made its first flight on January 25, 1957. The Jupiter, developed by Chrysler for the Army, but assigned to the Air Force, made sixty-five research and development launches in succeeding years, beginning in May 1957. In that same year, on June 11, the Air Force launched its first Intercontinental Ballistic Missile (ICBM), the Atlas, made by General Dynamics. The Cape was as busy as Times Square at midday, or Louisville on Derby Day.

One of the men who went to Florida to work on the Atlas, Wallace Dal Santo of Chicago, had taken a position in radar with General Electric in Utica, New York, on completing his bachelor of science degree at St. Joseph's College in Rensselaer, Indiana. In 1957 he moved his family to Titusville. He worked with a General Electric team that controlled the launch, the trajectory, and the termination of the flight. Wallace had the enthusiasm of the average space worker. "We could measure rates of pitch, swerve, velocity, and roll," Wallace recalled. "This told us where the missile was at the time. A Burroughs computer compiled all this data, did mathematical computations with it, and measured the information coming in. This helped us decide how we were to adjust the course of the missile. It told us when to cut the engines off so we could hit the prescribed target."[2] Then, with admiration in his voice, Wallace went on: "That Burroughs computer, the first transistorized computer, now holds an honored place in the Museum of Science and Industry at the Smithsonian Institution."[3] While Wallace and his fellows on the Atlas were rejoicing, the Soviets were exuberant. After eight unsuccessful launches of the Soviet R-7 between May and August 1957, on August 20, 1957, an R-7 traveled the full length of its range, splashing into the Pacific four thousand miles from the pad. Less than three weeks later, another rocket repeated this performance.

Soviet Premier Nikita Khrushchev, who witnessed the September launch, gave Sergei Korolev, a fellow Ukrainian and Russia's leading space expert, approval to move to a higher goal. While the R-7 was capable of lifting a ton, the Korolev team settled on a 184-pound satellite, called Sputnik, or fellow traveler. Outfitted with only a radio transmitter, this silver ball twenty-three inches in diameter, was to announce its presence in space.

The Soviets Shock the World

On October 4 of that rocket-happy year of 1957, the Soviets shocked the world with the successful launch of the first man-made satellite, Sputnik 1. To many in the free world, the Cold War had ended. The *Hot War* had begun. To those apprehensive of anything Communist, the Russians had waved a red flag. Many paused, reflected, and held judgment. But a few skeptics, aware that the Soviets had been taking credit for every invention from the safety pin to the jukebox, looked on the Russian claim as Marxist propaganda. An American newspaperman, for instance, interviewed Dr. G. A. Tokaty, a Russian aerodynamics expert, then head of the Department of Aeronautics and Space Technology in London, shortly after a successful Russian venture in space. When Tokaty refused to give "Yes" and "No" answers, the correspondent countered that it may well prove to be "another Russian hoax."[4] The basic misunderstanding stemmed from the lack of information the average American had on Russian rocket development. The nation knew of German advances and attributed any effort the Soviets made to German scientists and engineers working in Russia. A nationally known quipster had remarked to an approving audience: "Our German scientists could outdo their German scientists." That was true, but the Russians welcomed few Germans. Helmut Gröttrup was the only one of prominence. At first, they worked in East Germany. Later, relocated to Russia, they were separated and were never allowed to work as a team. *Our* German spacemen had to match talented Russian spacemen.

These men were Ukrainians, the already mentioned Sergei Korolev, and Valentin J. Lushko. Held back by Stalin, they moved ahead dramatically with the advent of Kruschev.

While the Pentagon and President Eisenhower played down the significance of Sputnik, many social observers wondered if it was one of those scientific breakthroughs that altered the balance of power. The average American was perhaps most concerned because someone else was excelling in technology—an area wherein the U.S. was accustomed to lead. It looked like a lovely star, not a danger-filled threat, as it passed over St. Louis at 7:10 on that October morning, moving northeast to southwest. Going in the opposite direction from midland tornadoes, it seemed as peaceful as it was beautiful. It was a memorable vision. The Russian dog, Laika, went up on November 3, 1957. Rockets could drop nuclear weapons on U.S. targets. The Soviets had declared a space war.

Less than four months later, on January 31, 1958, after the first Vanguard orbital launch attempt by U.S. Navy/Martin engineers had collapsed the previous month, the rocket team of Huntsville launched Juno. It carried into orbit the nation's first satellite, Explorer 1. At the same time, the team began studies of a booster ten times more powerful than the 150,000-pound thrust of Jupiter, that could put weather and communications satellites into orbit around the earth or propel a space probe out of the earth's orbit. But for the moment, the Huntsville team faced an uncertain future. The previous fall the Secretary of Defense, Charles Wilson, had assigned the responsibility for all intermediate and long-range missiles to the Air Force. What was the Army going to do? If it was going to stay in the big rocket business, it would have to find new tasks for its missile team at Huntsville. And so, General Medaris, Commander of the Army Ballistic Missile Agency, set his sights on a new super rocket that would win the name *Saturn* and explore space. He was farseeing.

Slightly less than a year had gone by since Sputnik 1 gave a new word to our vocabulary and a shock to the American psyche. If a space race was on, Americans wanted to take part—even if we had to start a lap behind. We'd get in the race and win it. With a cheering public, Lyndon B. Johnson, the Senate Majority Leader, pushed the Aeronautics and Space Act through Congress in 1958. Under its authority, President Eisenhower set up the National Aeronautics and Space Administration on October 1. He transferred the armed services' non-military space activities to the new civilian agency. NASA inherited all of the National Advisory Committee for Aeronautics (NACA) facilities; the Langley Laboratory and Wallops Test Station in Virginia; the Lewis Laboratory in Ohio; and the Ames Laboratory and Muroc Flight Station in California. Congress also authorized the construction of a third major laboratory, the Goddard Space Flight Center in Greenbelt, Maryland, that opened in 1961. President Eisenhower transferred to NASA all non-military space projects.

NASA had a mission, the exploration of space. The Huntsville team had missiles. Mission and missiles fitted perfectly. The following year NASA received a vital asset: the Army team of former German V-2 experts who were working with plans for *Saturn*, a large rocket. Assigned the task of manned space flight, NASA drew up a *Ten Year Plan of Space Exploration*. Revealed to Congress early in 1960, it called for nearly 260 varied launches during the next decade with a manned flight to the moon after 1970. The House Committee on Science and Astronautics considered it a good program but wanted NASA to move faster. NASA's immediate goal was the successful orbiting of a man aboard a Mercury spacecraft.

McDonnell Aircraft Corporation of St. Louis won the contract to build the Mercury capsule. This should have surprised no one. James McDonnell always looked ahead. Back in the early 1940s he had shown interest in jet propulsion and eventually designed and built what became *Phantom I*, the first carrier-based jet aircraft. This made his little firm a big one. With equal foresight, even before Sputnik 1, he had anticipated space flight and put forty-five engineers in space capsule research. One of these, Data Analyst Grattan Murphy, came to look on his work on the Mercury capsule as the most memorable moment in his entire work career. By the time NASA called on his firm, James McDonnell had invested $900,000 in Mercury style studies. It paid off.[5]

McDonnell Aircraft Corporation of St. Louis sought volunteers to go to Florida to work on Mercury. Joseph Szofran, one of the twenty-five men who went south at the time, remembers that Titusville had few available residences. Further, the causeway might close during the morning rush hour. Since the barges and boats had priority, the space men might wait a half hour to get over the Indian River. Some developers in the Cocoa-Rockledge area had sent representatives to St. Louis to tell people of the possibility of finding homes in that area. At that time, workers were coming and going, creating an active real estate market.

A large St. Louis contingent of technicians from the McDonnell Flight Test Department arrived in July 1960. They had experience with aircraft. Warned of the difficulty of getting parts and replacements at the beginning, many of them brought their own tools and whatever aircraft fasteners they could find. Every nut or bolt in the program needed to be approved and had to be made of extremely hard metal that could withstand the heat. The newcomers from St. Louis worked at the consoles on the Cape with the engineers in Hangar 6, and with the Redstone people. Employed by McDonnell, they worked for NASA at the time.

Under the direction of G. Merritt Preston, chief of NASA's Preflight Operations Division, the men felt the wonderment of doing something new. They were convinced that Mercury was going to be a success and had praise for Preston's management. At that time, however, no one talked about a further stage of the program. When the Mercury program ended, many presumed, the McDonnell people would return to St. Louis.

"When we first arrived," Szofran remembered, "we thought we could take a completed spacecraft from St. Louis, do some minor preparation to it, put it on the launch pad, and launch it. It sounded simple. But there were so many things that you just couldn't think of, never having done them before, that crept into programs. We found that we should do things in St. Louis that they were doing in Brevard, and things they were doing in Brevard that they should have been doing in St. Louis. It took a while to standardize these procedures."[6]

At the Cape, all booster stages in a spacecraft launch system first came together. Even though the manufacturers thoroughly checked and tested those components, engineers at the Cape had to make sure they were put together properly and worked. One of the Cape's main tasks was to check completely every system of the finished vehicle. If two vehicle components built by companies in distant states failed to work together as a system, the men at the Cape had to find out why and get them fixed.

No one had ever sent men into space. There were no precedents, no veterans to tell of similar experiences and offer solutions. The original cone-shaped design of the spacecraft made repair difficult. The packages of equipment were designed to fit one after another in a row. If salt-water or salt-air corrosion caused a problem, the space-workers had to remove much equipment to get at the affected part. To replace a battery along the side of the astronauts' console took about eight hours because of the many elements that had to be removed. To remove a misplaced washer took all available ingenuity and about eleven hours of effort.

All the while, the Cape Canaveral skyline already had distinctive features. Towering gantries rose along what came to be called the Intercontinental Ballistic Missiles Row or the ICBM Row in Cape parlance. Various missiles had certain similarities in ground needs and operational requirements. Each required an assembly and checkout building; transport from assembly area to launch complex; a launch pad; a gantry service tower; a blockhouse for on-site command in control of the launch; and a network of power, fuel, and communications links that would bring it to life.

Over a long stretch the complexes resembled each other. Igloo-shaped blockhouses stood 200 yards from the pads and looked like the pillboxes of World War II. They provided protection for the launch crew and the controls and instrumentation. In the case of Complexes 11, 12, 13, and 14, designed for the Atlas Intercontinental Ballistic Missiles, the inside walls of the twelve-sided domed structures were two yards thick at the base with fourteen yards of sand around them. Besides the blockhouse or launch control center, the essential features of a fixed pad complex included a concrete or steel pedestal on which to erect and launch the vehicle, a steel umbilical tower to provide fluid and electrical connections to the vehicle, a flame-deflector, and a mobile service structure that moved around the vehicle so ground crews on platforms could check and test various components. Other features of the complex included an operation support building, storage facilities for kerosene and liquid oxygen, a tunnel for instrumentation and control cables, roads, camera sites, utility services, and security. All this material had been shipped in from elsewhere, and the roads were narrow.

At that time highway U.S. 1 was a two-lane road to Melbourne. North Cape Road was narrow, lacking in shoulders, and hazardous. Atlas worker, Wallace Dal Santo, a colleague, George McClellan, and a few friends collected 15,000 signatures during 1959 and 1960 to have the road improved. The county commissioners were too busy to talk to them, so they went directly to Max Brewer, Commissioner of Roads under Governor Farris Bryant. Brewer knew the value of the space program and moved into action dramatically. He saw to road repairs. Grateful space-workers named the North Cape Road the Max Brewer Causeway.

The unconcern of the county commissioners may have reflected an attitude of certain old-timers who did not welcome the missile people and the Air Force. Involved in the citrus industry, they feared losing their cheap labor force, the orange pickers and packers, most of whom had long lived in the area. But the average resident of Brevard welcomed the newcomers.

Brevard Outraces the Country

During the years 1950 to 1960, when the Cape changed from an area of citrus groves, sand bars, and swamps to a major launch site, Brevard became the fastest growing county in the country. The average American county grew 19 percent during the decade. The state as a whole reached a high of 79 percent, 60 percent above the national average. Brevard soared from 23,653 to 111,435 individuals, an incredible increase of 371 percent, almost five times as high as the state's average and 19 times higher than the national average for counties. The median income had risen to $6,123, the highest in the state. Miami's Dade County stood second, with Jacksonville's Duval third.[7]

Matching the growth of population, manufacturing took its teen-age steps in Brevard in the 1950s. Only five firms still extant at the end of the century claimed origin before 1930: Couch Pump Company in Grant, Fabric Arts Studios in Palm Bay, Unifirst Corporation in Titusville, and R.S. Electronics and Harris Corporation in Melbourne. Only the last firm gave promise of future significance. Five other firms began in the 1940s: Precision Shops, Inc. in Titusville, Praxair Inc. in Mims, Helm Communication and Melbourne Sign in Melbourne, and Industrial Control in Rockledge. Eighteen more manufacturing firms opened during the 1950s. These firms produced boats, chemicals, signs, ironware, precision tools, storm doors, fiberglass molds, gates for parking lots, and other commodities.

Brevard County had no major city like Houston, where NASA headquarters planned to set up the Manned Spacecraft Center to develop Apollo spacecraft and train astronauts for space missions. Further, no one community dominated the Cape area as Huntsville, Alabama, was the focal point of the installations at the Marshall Space Flight Center. Instead, newcomers dispersed over a wide area that included all of Brevard County, Daytona Beach in Volusia County to the north, and Orlando in Orange County to the west.

A short distance south of Cape Canaveral, Cocoa Beach rapidly took a central role in the space program. Geography favored it. Many industrial contractors located there. Numerous motels and an excellent beach imparted a holiday atmosphere and made the town popular with tourists as well as space people. The area's nightlife centered there. The nation came to identify the space program with Cocoa Beach rather than with other communities in the vicinity. *Time* carried a sensationalized picture of activities at Cocoa Beach nightclubs on weekends and especially at launchings and splashdowns. Some space men celebrated at the Mousetrap Club. Bernard Surf was one of the area's better restaurants, and Ramon's was popular for its prime rib. Not unexpectedly, on such occasions, alcohol flowed freely as the padmen, like submariners on leave after a major venture at sea, drank numerous toasts in celebration of a successful launch.

But Cocoa Beach and the other space communities had another side. With the men busy at the Cape, their wives began to promote organizations for recreation and hobbies. Melbourne, Cocoa, and Cocoa Beach developed active theater and musical groups, including the Brevard Light Opera Association in Melbourne and the Brevard Civic Symphony in Cocoa. The Surfside Players at Cocoa Beach presented six plays a year. Later on, people in the growing service areas organized Rotary, Kiwanis, and other service clubs.

Melbourne, the metropolis of the county, was growing faster and was more stable than the county seat, Titusville. It was the home of Harris Corporation, the only firm in the county with a payroll of significance at the start of the space program. Harris grew considerably and kept pace with the activities on the Cape. It would continue to expand. The Florida Institute of Technology developed during those years. At first, the scientists and engineers at the Cape taught the classes at night. Gradually, the institute became a typical technical school with academically oriented professors teaching full-time. New churches, such as the Our Saviour that opened in 1956, began to supplant the Barn Theater and the bowling alley as part-time places of worship for the space-work families. Brevard Community College began in an abandoned apartment complex in Cocoa in 1960 and grew steadily.

All the while, Brevard County had no television station. As a result, the cities of Orlando and Daytona Beach influenced the region through their television facilities, even though they were respectively, forty and fifty miles distant.

Cocoa Beach had a charmed location for recreational activities. Even if one did not swim, a walk on the beach in the evening tended to soothe the spirit. Many found such walks rejuvenating: for example, a woman engineer who faced the realization that her male colleagues would not accept her professional credentials and on that day asked her to correct and retype their misspelled reports; or the design specialist who learned, after his family had dug its roots hickory-deep in Florida's sandy soil, that he had to return to the cold North or lose his job. While Cocoa Beach renewed the spirits of many, space-workers in Titusville lacked a convenient beach. After a day of worrying about the hatch on the spacecraft, an engineer of North American could not swim away his worries at rough but challenging Playalinda Beach. Just five miles from Titusville, it remained closed during the time of the flights.

Beach or no beach, Titusville had a promising future. One of the non-space employers to realize that was Don Boland, a Miami homebuilder. Invited to visit the area in November 1956, he saw its possibilities and hoped to relocate there. But his wife, Chris, was reluctant to move. As a girl of eight, she had been forced to flee with her family from her home in the Ukraine to escape Stalinist tyranny. Miami had become heaven for her. When her husband pointed out the prospects in North Brevard, she reluctantly left Miami and found happiness in a spacious home on Indian River.

A tiny percentage of the residents of the four main communities of Brevard County had been born there. According to the 1960 Census, roughly one-fourth of the newcomers came from each of these categories: villages with less than 5,000 residents; towns with between 5,000 and 25,000; cities with from 25,001 to 100,000; and cities over 100,000. Industrial firms transferred 13 percent of the newcomers from plants in others areas; 25 percent freely accepted Florida jobs with a firm they already worked for; and slightly over 25 percent sought better economic opportunities by coming to the area to seek employment. Close to 20 percent came from other counties of Florida; 35 to 40 percent came from other southern states; and 15 to 20 percent moved down from the North. Thus, over half of the early space-workers were southerners.

After 1960, the space-workers came from other areas, especially big cities. McDonnell sent men and women from St. Louis to work on Mercury and Gemini. Boeing personnel came from Seattle, Grumman from Long Island, North American from Downey, California, and Douglas from Santa Monica and Huntington Beach, in that same state. A number of individuals, including Walter Cooney, Jack Tobin, and Ray Cerrato, came from Parks College of Aeronautical Technology, a school Dr. von Braun often visited in the hope of recruiting qualified young engineers.

The space people brought their home loyalties with them, including their favorite sports teams. The Boeing workers cheered for the Seattle Seahawks. Grumman workers rose and fell in spirit with the New York Mets. Douglas workers agonized over each Los Angeles Rams' loss. The McDonnell crew applauded the St. Louis Cardinals' victory over the New York Yankees in the World Series. On the college level, the big game split the Huntsville group between "Bear" Bryant's Alabamans and ever-dangerous underdog, Auburn. But all Alabamans joined together on New Year's Day when the Crimson Tide tried to halt the Fighting Irish of Notre Dame. Still later, the steadily improving Miami Dolphins surged as the favorite team of otherwise uncommitted sports fans.

As the 1960 presidential campaign approached, the Democrats had many able candidates to challenge Vice President Richard Nixon: Senators Hubert Humphrey of Minnesota, Stuart Symington of Missouri, and Lyndon Johnson of Texas. The Texan's serious concern for space exploration made him believe he could ride that issue into the White House. Coming up on the outside was a young senator from Massachusetts, John Fitzgerald Kennedy. He had other priorities than space, but he was nominated. When he debated with Vice President Nixon, Senator Kennedy's wit and charm won over many American people. They elected him president with Lyndon Johnson as his vice president.

Even though neither Kennedy nor Nixon had given major attention to space exploration in the debates, on July 28 of that summer, the people of NASA announced a new manned space flight program. Called Apollo, it planned to put three astronauts into a long-enduring orbit around the earth or into a flight around the moon. The timing of the announcement caused jitters. The next day, at Cape Canaveral, NASA's first Mercury-Atlas (MA-1) disintegrated and fell into the ocean fifty-eight seconds after takeoff. This disaster ushered in a bleak four months during which the test rocket Little Joe 5 joined the MA-1 in the ocean, and the first Mercury-Redstone lifted a fraction of an inch and settled back on its launch pad. The last failure, on November 21, marked the absolute nadir of morale for the engineers working on Mercury.

The people at NASA's new headquarters in Washington, coping with financial and administrative problems and facing a change of administration after the election of John F. Kennedy as President, were only a little less dispirited than the workers in the field. Was Kennedy as interested in space as Vice President Johnson clearly was? The fledgling space agency had its problems, but it also had a great asset: the growing interest of the American people in space ventures.

On April 12, 1961, cosmonaut Yuri Gagarin, a twenty-seven-year-old Russian Air Force pilot, circled the globe. The Soviets had put up the Berlin Wall and Premier Nikita Khrushchev threatened to "bury" the U.S. Locked in confrontation of prestige with the Soviets in Cuba, in Berlin, and now in space, America had to answer these threats. President Kennedy asked the scientific community to come up with a dramatic response. Could they find a way to desalt ocean water? What of a program to send men to the moon and bring them back? NASA held it was feasible. The scientific community had no alternative, but hesitated to support manned launches into space, even after the chimpanzee, Ham, came back to earth safely in 1961.

Astronaut Alan Shepard and the padmen at the Cape helped to move the question from discussion to action. The time came for the first *manned* flight by an American in a Mercury spacecraft on a Redstone rocket. The von Braun team at Huntsville had developed the booster. The spacecraft owed its conceptual design to the Space Task Group at Langley Research Center, Hampton, Virginia, under the direction of Robert R. Gilruth. A veteran of the research center from the days of NASA's predecessor, the National Advisory Committee for Aeronautics (NACA), Gilruth had demonstrated his ability working with test pilots and moving into supersonic flight research. The Mercury spacecraft owed its manned space flight concept to Gilruth's engineers, but its advanced and detailed design were a product of the forward thinking of McDonnell Aircraft's James S. McDonnell. When the Russians astonished the world in 1957 by putting up the first man-made satellite, McDonnell had forty-five engineers at work for five months researching the basic problems of manned space flight. When McDonnell was awarded the contract to design, build, test, and deliver the Mercury spacecraft within a short time period, Gilruth's engineers were there to oversee that work. John Yardley headed the McDonnell team.

With thousands watching from every corner of Brevard County and as far away as the beaches of St. Augustine and Boca Raton, and with millions of hopeful Americans at their television sets, America's pioneer spaceman, Alan Shepard, soared beyond earth's atmosphere in his Freedom 7 for fifteen minutes on May 5, 1961. He controlled the spacecraft while in flight and landed safely 302 miles downrange.

Accustomed as Americans became to many successful launches over the succeeding years, the shortness of time between the steady failures at the Cape and the success of the first manned flight, when viewed in the documentary *Thrust Into Space*, staggers the viewer. Shepard sat calmly on a giant bomb and triumphed. The first astronauts deserved accolades as supermen.

The moment called for a dramatic move.

2

ANSWERING THE CALL
OF PRESIDENT KENNEDY

JFK Challenges His Fellow Americans

Convinced it was necessary to show the world what America could do, President Kennedy told Congress on May 25, 1961:

Now is the time to take longer strides—time for a great new American enterprise—time for this nation to take a clearly leading role in space achievement, which in many ways may hold the key to our future on earth. I believe that this nation should commit itself to achieving the goal before this decade is out of landing a man on the moon and returning him safely to earth. No single space project of this period will be more exciting or more impressive to mankind, or more important to the long-range exploration of space, and that will be so difficult or expensive to accomplish—in the very real sense, it will not be one man going to the Moon, it will be an entire nation, for all of us must work together to put him there.

Like President Lincoln at the firing on Fort Sumter and President Roosevelt at the attack on Pearl Harbor, President Kennedy had little trouble in moving the nation to react. Brevard County certainly did.

This challenge, and the confidence it showed that the men and women of NASA could bring it off, swept the Cape like a spring breeze off the Atlantic. Before this, many Americans had been skeptical of space exploration, even when the Russians launched their vehicle. Others looked upon rocketeers as grown-up children who liked to toy with Paul Bunyan–sized firecrackers. Now they responded to the challenge of the young President. The people were behind him and the men and women of NASA. It was not to be *easy*. But it *was* to be.

To meet the Kennedy challenge, NASA had to answer two major questions promptly. The first asked how to send men to the Moon—by direct flight, by earth-orbital rendezvous, or by lunar-orbital rendezvous? The second pondered where best to launch the astronauts on their way—from the Cape or from another possible site, such as White Sands, New Mexico, the southern tip of Hawaii's big island, or Cumberland Island, Georgia?

As to the first question, Wernher von Braun recommended putting a space station in orbit and having the moon vehicle assembled there—a good plan, but one that seemed impossible within the time frame. Robert R. Gilruth, head of the Space Task Group at Langley Research Center in Virginia, favored direct flight to the Moon. The two leaders called joint meetings to foster understanding. One of the researchers on the Space Task Group, Dr. John C. Houbolt, strongly advocated a lunar-orbital rendezvous. A persistent man, Houbolt advised sending the Apollo to orbit the moon and then having a lunar-landing vehicle go to the moon and return to the command module for the return trip. The men of the Cape dubbed this last plan the "Santa Maria-rowboat way." They pointed out that Columbus did not float the Santa Maria to the sands of San Salvador, but anchored the vessel in the bay and then sent a skiff to the shore. Houbolt gradually made others listen.

In June 1962, a team under Joseph F. Shea, Deputy to D. Brainerd Holmes, director of the Office of Manned Space Flight, met with von Braun for an all-day session. Late in the day, to the surprise and dismay of his staff, von Braun supported the lunar-orbital rendezvous as offering the best chance of completing the task within the decade. That marked a big step forward, but discussion continued. President Kennedy joined in the appraisal when he visited Huntsville in September 1962. Finally, in December, NASA decided on Houbolt's plan, the lunar-orbital rendezvous.

In his book *Stages to Saturn*, historian Roger E. Bilstein praised the planners for their open discussion, but insisted: "This is not to say that differences of opinion were always easily and quickly adjusted."[1]

The second question, where to locate the Moonport, proved less easy to answer. A joint NASA-Air Force team, cochaired by Dr. Kurt Debus, director of NASA's Missile Firing Laboratory, and Major General Leighton Davis, Commander of Patrick Air Force Base, considered eight sites: Cape Canaveral; offshore from Cape Canaveral; Mayaguana Island in the Bahamas; Cumberland Island, Georgia; a mainland site near Brownsville, Texas; White Sands Missile Range in New Mexico; Christmas Island in the mid-Pacific south of Hawaii; and South Point on the island of Hawaii.

Several theoreticians urged an equatorial launch site, such as Christmas Island south of Hawaii in the Pacific Ocean. It stood on the equator and thus could take full advantage of the earth's rotation, but the logistics of bringing all of the supplies there militated against that one element of superiority. Further, recruiting qualified scientists for any length of time in such a remote area baffled the experts. A few analysts suggested Kalae, the southern point of the big island in Hawaii. It had the advantage of being on American soil. Again, the difficulty of bringing supplies and people to that distant place shouted "No!"

White Sands, New Mexico, was likely to cost least to develop and operate, but the area was landlocked. The lack of water transport virtually dictated construction of the space vehicle, assembly plants, and firing stands near White Sands. There'd be no shipping of components by water from elsewhere. Further, errant missiles could threaten many areas of Latin America. The missile-men had already moved from there for that reason. Any site in Texas required flights over Latin America and rated a hazardous recommendation. That left two places—central Florida's east coast or Cumberland Island, Georgia.

Like the Cape, Cumberland Island offered undeveloped land, railroad facilities, a coastal waterway, and port facilities. It stretched north and south for twenty miles, and was slightly over three miles across at its widest point. The intracoastal waterway, tidal flats, and saltwater marshes separated it from the Georgia mainland. Deep-water docks along the intracoastal waterway provided access to low cost water transportation. Government-owned King's Bay ammunition facility stood close at hand. Proponents could rightly look to relatively low real estate costs, even though to the north, expensive resorts flourished. Further, it had one feature of its own, proximity to a large city, Jacksonville, Florida.

Not much farther from Cumberland Island than Melbourne and Titusville were from Cape Canaveral, the city of Jacksonville offered universities, schools, churches, theaters, museums, recreational opportunities, occasional major sports activities, department stores, shopping malls, and needed services. Its metropolitan area stretched north within easy range of the people working on Cumberland Island.

While it had some of the Cape's advantages—railroads, flight patterns over the ocean, limited booster impact, accessibility to deep-water transport, and mainland location—it had one big disadvantage: it was not on the line already established to track the flights. Instead, it required a dogleg path to get on the line that stretched directly from the

Cape area to the Caribbean, South Atlantic and, ultimately, the Indian Ocean. Further, it was the last unspoiled island preserve on the east coast, kept that way in modern times by the Andrew Carnegie family and later by the National Park Service. Perhaps for this reason, influential Senator Richard Brevard Russell of Georgia, head of the Senate Armed Forces Committee, failed to push the location in his state.

Even before the submission of this report, Dr. Debus knew that the Air Force looked on space as its area. He had misgivings about NASA's grip on the purse strings in the event the Moonport was located within the Air Force's sphere of influence. Milton W. Rosen, acting director of Launch Vehicle Programs, called for a more complete study of Cumberland Island before a final decision in favor of the Canaveral area. Rosen wrote, "At Cumberland . . . there is an opportunity, one which we should not lose, to operate in a much simpler and more effective and less time-consuming manner. At Cumberland there could be, at the beginning at least, essentially one project directed toward a single, major objective. The newness of Cumberland would be an asset. Both White Sands and Canaveral had simpler and more direct, less time-consuming procedures in their early days."[2] Cumberland might bring back these days.

Further, since the Air Force had no base in the area, no rivalry or conflict of interest would arise. Several workers noted a practical fact that persons living in the northern suburbs of Jacksonville could drive to work on Cumberland Island through less traffic than employees faced at Cape Canaveral. The cost of duplicating instruments paled in contrast to the investment at either site. Further, labor conditions were bad in Florida. Since it had never been an industrial state, it lacked skilled workers in various categories and had none at all in others. NASA and its contractors had to call in engineers, scientists, and other experts from various parts of the country or train local people in a wide variety of skills. Along with the men, manufactured goods had to come from elsewhere—nuts, bolts, copper wire, power instrumentation, cables, transformers, circuit breakers, and generators, to list a few. Could the Canaveral area serve the needs of the countless families NASA was bringing for the moon launches?

Canaveral *Si*, Cumberland *No*

Shortly after the Shepard flight, Raul Ernesto Reyes came to Florida to conduct tests on Mercury flights.[3] As a young boy in El Paso in 1947, he had become aware of rocketry when he heard the thud of the errant missile that landed across the Rio Grande in Juarez, Mexico. Over his college years at New Mexico State his interest grew. In 1960, he served with the Air Force in Dayton. After he married in 1961, he went to work for NASA in Houston.

When Ernie arrived in Brevard in 1963, the county had no major hospitals, libraries, schools, sidewalks, or shopping centers. It lacked houses, stores, schools, and recreational facilities for new residents. It had few grocers, butchers, hardware merchants, cobblers, watch repairers, barbers, hairdressers or seamstresses. When your shoes needed repair, you drove a two-lane road fifty miles to Orlando to get to a cobbler shop or to buy a new pair. If your watch stopped, you sent it to Tampa to get it fixed. If you needed a book for permanent use, you ordered it by mail. If you lived in Merritt Island and wanted to worship God on Sunday, you went to the Barn Theater for services, no matter what your denomination. Brevard was the fastest growing county in the country and had scarcely kept up with pre-Apollo growth.

As a result, some NASA chiefs doubted the area's ability to support the Apollo program, but Senator Robert Kerr of Oklahoma, the powerful head of the Senate Committee on Aeronautical and Space Sciences, stressed several factors that pointed out the value of working where they were, especially the tracking network that stretched into the Indian Ocean. If NASA tried to start from scratch in another area, he believed that this one aspect of the program outweighed other considerations. NASA could plan efficiently for future expansion in the Canaveral area, Senator Kerr insisted, and he urged the agency to plan for long-term use.

Had human considerations outweighed the financial and scientific, Cumberland could well have won. And this book, if written, could not take pride in being part of a series on the development of Florida. But NASA chose the Cape, and this book came to be.

Six days after Senator Kerr's intervention, on August 24, 1961, NASA headquarters announced plans to acquire approximately 125 square miles of land, largely on Merritt Island, north and west of the Cape Canaveral launch area, for manned lunar flights. While some observers felt that the deciding factor was financial, and others that it was the tracking line, Colonel Asa Gibbs of the NASA support office held that the deciding factor was safety.[4]

In September the Army Corps of Engineers undertook the land acquisition. By the time all owners had listed their holdings, the number of tracts reached 440. Three out of four owners were absentee. Over half lived outside of Florida. The corps hired experienced land appraisers from Florida firms and issued a booklet to explain the procedures. The corps intended to identify the owner, map the land, and describe it briefly. The appraiser could evaluate each tract, then the corps would negotiate with the owner. Generally, negotiations proved successful, and a representative closed the deal. In some instances, however, negotiations broke down and the government had to begin condemnation proceedings.

While not involving a great number of people, the disruption had its poignant elements, as do all such transfers. One family had come down from Savannah, Georgia, a few years before and purchased a small estate near Happy Lagoon, about a mile and a half north of where the assembly building was to arise. Husband and wife had come to cherish their new location. The Corps of Engineers assured them that, if they purchased similar land north of Haulover Canal, they need never worry about moving again. That proved wrong. Later, NASA had to reassess its needs and decided to expand farther north, and the harassed couple finally moved to Orlando. Some individuals moved their houses to the mainland or to the south end of Merritt Island, and one of these buildings became in 1958 the original St. Theresa's Church in Titusville.

On the same day that NASA announced its intention of obtaining land on Merritt Island, it signed the Webb-Gilpatrick Agreement with the Department of Defense (DOD). This compact set guidelines for managing and funding the manned lunar landing program. Taking its name from James E. Webb, whom President Kennedy had just appointed to head NASA, and Deputy Secretary of Defense Roswell L. Gilpatrick, it set down three preliminary considerations. Both NASA and the DOD recognized the great impact of the manned lunar landing program on the Atlantic Missile Range. In the national interest, the two agencies were to pool their resources and make the most effective use of the facilities and services so that the traditional relationship between range users and range operators might continue. In this area of east central Florida, denuded now of residences, NASA planned a mighty launch center for Apollo with an amount of major construction never seen before in Florida. And it planned to do this in a region with few industries, no strong unions, and few skilled workers.

In 1961, Walter Williams and Christopher C. Kraft, Jr., Houston's mission and flight directors, came to the conclusion that, in view of the longer and more complex missions of Gemini and Apollo, something larger and better equipped than the Mercury Control Center at the Cape was needed.[5] They concurred that the center should be on Texas soil. On July 20, 1962, NASA administrator James Webb authorized the setting up of a manned space center near Houston. As a result, the American people at their TV sets began to identify space activities with Houston, not with the Cape. Many Floridians looked on the Houston installation as unnecessary. They thought they could handle manned space flight along with their other duties and looked on the Texas decision as more political than scientific.

The ruling favoring lunar orbit put the members of Gilruth's Space Task Group and other engineers and scientists at Houston, now the Manned Spacecraft Center, hard at work on the spacecraft for Apollo. The choice of a lunar-orbital rendezvous called for two vehicles whose construction the Manned Spacecraft Center was supervising. North American Aviation in Downey, California, was building the command ship, consisting of a command module and a supportive service module to carry the crew to lunar orbit and back to earth. At the other side of the continent, on Long Island, Grumman Aircraft engineers worked on the lunar module. This spidery-looking spacecraft was to set two men down on the moon's surface and return them to the command module waiting in lunar orbit for the return trip to earth. It was a unique and costly vehicle with no other foreseeable use.

Also in that eventful year in space history, Soviet cosmonauts indirectly aided America's progress beyond earth. In flights of Vostok 3 and

Vostok 4 they showed that men could live in space for a long period of time and could rendezvous there.

Building the Spaceport

No sooner had the spacemen chosen the lunar-orbital rendezvous and picked the Cape over Cumberland Island and all other sites as the place to launch from, than a third question faced the men of Apollo: Should the team assemble the spacecraft on launch pads as it had done with Mercury and Gemini? New problems affected Apollo—the size and complexity of the Saturn V, the need for unprecedented reliability with flexible launch rates, and a short recovery time between launches from the same pad. Harvey Pierce of Connell and Associates, engineering consultants on launch pads 34 and 37, urged the Mercury approach, as did the widely respected NASA design engineer James Deese, of the Facilities Office, who offered criteria for L-39 and later for an enclosed assembly building.[6]

Dr. Kurt Debus, director of NASA's Launch Operations Center, insisted on assembling the vehicle in a large building and then transporting the assembled Apollo-Saturn to the new launch pad to be set up near the ocean, north of Cape Canaveral. Russ Gramer of the Quality Surveillance Division recalled that no real discussion ensued on the mobile concept as with the rendezvous issue. But long discussion was involved with another major problem: how to move the moon vehicle from the assembly building to the pad—by waterway, rail, or pike? And if by pike, what surface should it have to transport a Washington Monument-size apparatus from assembly building to launch pad?

The decision to erect the largest building in the world as a place to stack the Apollo-Saturn spacecraft might have daunted even the road-builders of ancient Rome or the planners of the pyramids of Egypt. It overreached the imagination of the average person. It had to be tall enough to allow for stacking the 360-foot-high Saturn V on top of its 46-foot-high loadable launch platform and large enough to enclose an area of 129 million cubic feet, cover some eight acres, and be able to meet severe design loads. Bridge cranes had to span 147 feet and lift 110 tons to a height of 197 feet. To be the largest building in Florida, and for a short time in the entire country, it had to have high checkout areas, each big enough to handle all stages of the Saturn V and the spacecraft in a stacked position, ready for launch. The Vertical Assembly Building, or VAB, needed an extra strong wall to withstand both hurricane winds and the acoustics pressure from the thundering Saturn V-5 engines.

The building had to face fierce winds up to 125 miles per hour and waves and acoustical pressures from launches three and a half miles away. In an area with more swamp than dry land, more mosquitoes than mockingbirds, and more mud than base rock, the men of NASA were erecting a structure on whose top the Yankee Stadium could fit. Its doors reached higher than any building in Florida, it had to be huge because of its assigned task, and it had to be kept from sinking slowly into the soft southern mud. Eventually, it cost $112 million.

In August 1962, a committee of the Launch Operations Center asked the U.S. Army Corps of Engineers to select an architectural-engineering firm to complete the criteria for the VAB. Corps members from the various areas submitted a list of five firms. The idea of a joint venture had come early in 1962, when Max Urbahn, head of an architectural firm, and Anton Tedesco, a structural expert with Roberts and Schaefer, Inc., who had earlier engaged in similar projects, had discussed the possibility of designing a lunar launch center in Florida. Both had worked on joint ventures before. They invited A. Wilson Knecht of the electrical firm of Seelye, Stevenson, Value, and Knecht, and Philip C. Rutledge of Moran, Proctor, Meuser, and Rutledge, a firm that designed foundations, to join them. All these firms had experience in the area of concern. The chief of the Corps of Engineers selected a combine made up of these firms that came to be called URSAM, a name derived from the first name of each of the companies. They set to work.

At the same time, dredgers moved nine million cubic yards of sand and shell to form the access channel for the spacecraft barges to carry the missile components on the last stage from Tennessee and Louisiana to the VAB. Most of the sand and shell went to build the three and a half-mile crawlerway that stretched from the VAB to launch pads 39A and 39B. In short, they were digging a canal to bring components of the Saturn V from the Atlantic, and parallel to it, a roadway unlike any other in the nation, to transport the assembled Saturn missile vertically the three and a half miles to the launch pad near the ocean, on a vehicle of as yet uncertain size and shape.

The crawler-transporter truly merited its nickname, "The Texas Tractor." It could have won a blue ribbon for sheer ugliness. From a distance, it looked like a steel sandwich, held up at the corners by four vehicles that resembled World War I tanks. Each crawler was larger than a baseball infield and weighed about 5,000 tons. Two 2,750-horsepower diesel engines powered sixteen tractor motors that moved the four double-tracked treads. Each tread had fifty-seven "shoes," and each shoe, measuring almost a foot by seven and one-half feet, weighed close to 2,000 pounds. Because of their importance and their cost, they were referred to as "Them Golden Slippers," recalling the line of the song, "Golden slippers I'm going to wear, to walk that golden street." The crawlerway was such a street.

While men had moved two- and three-story houses often enough, even some from Merritt Island to the mainland of Florida, no one had ever before moved a skyscraper. Yet that is what the mobile concept called for, or at least an Apollo-Saturn vehicle the size of a skyscraper. The problem was compounded by Merritt Island's marshy terrain and high winds. The combined weight of a crawler-transporter, a mobile launcher, and an Apollo-Saturn exceeded seven million pounds. Would the subsoil hold that? C. Q. Stewart of the Mechanical Engineering Division commented on this problem in a memorandum of August 1, 1962. He suggested exploratory borings, advised against any type of rigid surface as too prone to cracking, and recommended instead a topping of gravel or crushed stone. NASA heeded his advice. The surface of the roadway was Alabama river rock.

Unusual in itself, this amazing vehicle required unusual men to operate it. Imagine moving the Washington Monument three miles down Pennsylvania Avenue at six miles an hour, then keeping it erect while climbing the terrace in front of the Capitol. Driving the crawler was like braving the Los Angeles freeways in a slow-moving jalopy. One of the drivers, Lieutenant-Colonel Lee Verbeck, U.S.A.F., had piloted his B-24 Liberator on memorable flights over Nazi Germany before being shot down and enduring cold months in Stalag 1 on the shores of the Baltic. Lee stayed in the Air Force until 1956. He then went to Florida to work for Bendix and Boeing on the crawler-transporter.

A fellow crawler-driver, Fred Renaud, a native of Canada, had the physique of a defenseman in the National Hockey League. Fred worked with the missile program in Italy until the Cuban Missile Crisis brought a change. Transferred to Vandenberg Air Force Base in California, Fred worked for Bendix before reassignment to Florida for the crawler.

"I was on crawlers from their inception," Fred stated. "We had a crew of eighteen to twenty for the two crawlers. We learned from drawings and writing procedures, developed those procedures, tried them out, and more or less went through a year and a half of training on our own. We developed driving techniques and operating skills that everybody had to have for various functions on the crawler. The hydraulic room people had special talents."[7]

Lee Verbeck, Fred Renaud, and their fellow drivers were indeed unusual men. But at the spaceport, many groups of workers were out of the ordinary.

The mobile launcher, or launch umbilical tower, that rode the crawler down this road of river rock, along with the Apollo-Saturn, had three main features. Its base was a two-story platform 161 feet long and 131 feet wide, with a fifty-foot square opening in the middle for rocket exhaust. The launch vehicle itself, held erect by four countdown arms, stood on the platform, both on the journey from the VAB to the pads and on pads 39A or 39B. A square steel tower, roughly the size of the Apollo, stood beside it, surmounted by a hammerhead crane. Attached to the tower, swing arms of various sizes carried electric, pneumatic, and propellant lines to the space vehicle. The two-story launch platform housed computers connected with the Launch Control Center. Two high-speed elevators provided access to levels of the complicated apparatus.

With the mobile concept prevailing, the two octagonal pads at launch complex 39 were less complicated than those of earlier launch sites. The new stationary facilities included a terminal connection room that housed electronic equipment, and a single story environmental control room. The pad that rose forty feet above ground level had to support the weight of the Apollo-Saturn and the mobile structures.

Equally necessary, though less complicated than the crawler and the mobile launcher, the flame deflector stood directly under the moon vehicle at launch time. This wedge-shaped steel device, fifty feet wide, almost 100 feet long, and forty-three feet high, sent the fiery exhaust of the five first-stage engines along the flame trench.

At the same time, URSAM, the combine that planned the Vertical Assembly Building, also designed the Launch Control Center (LCC). This final building in the launching area, a four-story, monolithic, re-inforced concrete building, measuring 375 by 180 feet, made exten-sive use of pre-cast and pre-stressed elements. The architects wanted it to be symbolic. The VAB was to be the factory, and the LCC man's window for observing events projecting into the future. The building bore multi-level firing rooms, rectangular in shape, 90 feet in width, and 150 feet in length. Since many checkout requirements were still unknown, the planners emphasized flexibility, eliminated all columns, and provided removable floors. The design of the windows shut out the eternal sounds and pressures. The eight-inch-thick glass windows, with adjustable sun visors and special aluminum frames, faced the launch area. Infrared lamps outside the windows prevented fogging. The tinted, laminated windows that covered an area 78 feet long and 23 feet high filtered out heat and glare, permitting only 28 percent of the light to enter the room. Transparent glass, separating the viewing section from the rest of the firing room gave guests the feeling that they were part of the operation. The LCC brought URSAM the Architectural Award for the Industrial Design for 1965.

Various companies took part in designs and plans for the industrial area, including the Guided Missile Range Division of Pan American World Airways. This unit had worked at times with the Air Force on sev-eral projects. The Apollo mission support facilities included the head-quarters building, an operations and checkout building, an ordnance storage facility, a fluids test complex, and structures for shipping, supply, and receiving.

During all of this construction, while working men toiled and suf-fered from the elements, a few died in accidents. On June 4, 1964, workers were stacking concrete forms for the third-floor deck in the low bay area of the VAB. Apparently, the forms became overloaded and collapsed. Five men fell and were injured, two seriously. A month later, on July 2, 1964, Oscar Simmons, an employee of American Bridge and Iron Company, died in an accidental fall from the forty-sixth level of the VAB. On August 3, 1965, lightning killed Albert J. Traib on Pad 3 of launch complex 39.

The nearby passage of hurricanes Cleo in late August and Dora in early September 1964 caused an estimated $35,000 worth of damage but a delay of only three days. The editor of the *Spaceport News* reported the lack of major damage to NASA's facilities and dispensed credit widely, from the people who drew up the storm plans to the man who laid the last sand bag in place before Cleo swept in. The editor singled out Jim Jones, an engineer with the Instrumentation Division's Acoustic and Meteorological Section. The son of a clock salesman, who narrowly escaped danger in the 1926 Miami hurricane, Jones had a family-instilled wariness of big winds. A bachelor at the time, Jones had volunteered to ride out the storm alone in the huge launch control center in Complex 37, gathering weather data. From 10:00 a.m. on Thursday until relief came at 7:30 the next morning, he recorded winds that peaked at seventy miles per hour. The reluctant hero admitted that he had misgivings during his lonely vigil, even though he had thought that the launch control center looked like the safest place in the vicinity. The mild-mannered weatherman gained the name "Hurricane" Jones for his efforts.[8]

During those months, the word "integration" was an "in" expression at the spaceport, but far removed from the racial connotation it had elsewhere in America. At the spaceport it meant the working together of civilians and colonels, the older companies like North American and the new companies like McDonnell. It came readily among the ordinary people on the pad, who realized early on that everyone was important and all their work had value. Eventually, it had to come on all levels, and the biologist and physician had to sit down with the engineer and work out solutions to countless problems before man could reach the Moon.

What was the "state of the art" as of May 25, 1961, when President Kennedy challenged the nation? Since December 1957, when the first Vanguard orbital launch attempt had collapsed in flame before a television audience, the United States had tried to put twenty-five other scientific satellites into earth orbit; ten had been successful. America's space-probers had placed two meteorological satellites in orbit, and both had operated properly. Teams had launched two passive communications satellites, but only one had achieved orbit. Of the nine probes launched toward the Moon, none had hit its target, although three achieved limited success by returning scientific data during flight.

Russia had done better. Back in January 1959, the Soviets had sent up a spacecraft, Lunik 1, to fly by the moon. On the fourth day of that year, it had done just that, at a distance of less than 4,660 miles. A small object, hardly forty inches across, it blazed a trail and sent back information about the Moon's magnetic field—or rather the lack of it. Lunik 2 went up in September 1959. Scientists in Russia and England followed its signals until impact on the moon presumably destroyed it. Lunik 3, on October 10, moved around the moon and swung back earthward. It transmitted pictures to the world.

In the probe of Venus, also, Russians took the lead. They launched the Nira 1 on February 12, 1961. Men on earth lost contact after a few weeks when the probe reached a distance of 4,650,000 miles. Although Venus is the nearest planet to Earth, it is always at least a hundred times as far away as the moon. This fact alone made contact difficult enough. But probes had to maneuver through a complicated process. The spacecraft had to go in orbit around the earth, slow down, then swing inward toward the sun to meet the orbit of Venus at the proper time. Obviously, a journey to Venus took many weeks. That prospect lay far ahead.

The immediate question was this: would cosmonauts reach the moon before astronauts?

More Mercury and Vostok Flights

Other Mercury flights matched the pioneering Shepard during the space center's building years. Virgil I. "Gus" Grissom had followed with aplomb on July 21, 1961, but almost drowned before recovery at sea, when the hatch blew open prematurely. Russia followed with a successful launch on August 6 of that year. Major Gherman Titov circled the earth seventeen and one-half times and landed safely.

The space race was heating up. John Glenn sailed into orbit on February 20, 1962, the first American to do so. Astronaut Scott Carpenter made three orbits around the earth on March 24 of that year. A Soviet pair, Andrian Nikolayev and Pavel Popovich, went up in Vostok 3 and Vostok 4 on successive days. They could not rendezvous or dock, but did fly in tandem three or four miles apart. In the fall, Walter Schirra circled the earth six times. Gordon Cooper went up on May 15, 1963 and completed twenty-two orbits. Valery Bykovsky went up with Vostok 5 on June 14, 1963, and Valentina Tereshkova followed with Vostok 6 two days later. Valentina was a sport parachutist, not an experienced jet pilot. All American test pilots had been jet pilots, a requirement that irked some American scientists.

"The Vostok flights were impressive," historian Tom D. Crouch stated in his book *Aiming for the Stars*, "but Americans had recovered from their crisis of confidence."[9]

Mercury had turned to gold.

John Yardley, a stress engineer, managed the Mercury program for McDonnell. "He was an exceptional engineer," one of his colleagues stated, "and an exceptional manager for being an engineer. Generally, engineers aren't thought of as being great managers, but he was both." When Mercury developed, he became program manager and launch manager at the Cape. The program had succeeded beyond expectations.

By this time, spaceport workers deserved a moment of relaxation and a good laugh. They got both a month after Cooper's trip into space, when the *Spaceport News* for June 20, 1963 carried this startling headline: "The Duchy of Grand Fenwick Takes Over the Space Race Lead."[10] The article told of the premiere at the Cape Colony Inn on the previous Friday of a British movie, a space satire called *Mouse on the Moon*. Distributed by United Artists, the movie was a sequel to the popular *The Mouse That Roared* of several years earlier.

The Mouse That Roared centered on the attempt of the Duchy of Grand Fenwick, a mythical principality near the Swiss-French border, to wage an unsuccessful war against the United States in the hope that the United States might pour millions of dollars into the country for rebuilding. Surprisingly, the war turns out to be a huge success for the Grand Fenwick Expeditionary Force. It captures a professor at Columbia University, a native of Grand Fenwick, who has invented the "bomb to end all bombs." By threatening to use the bomb on all the major nations of the world, Grand Fenwick brings universal peace.

In the sequel, *Mouse on the Moon*, Grand Fenwick, faced again with a disaster in its main industry, wine-making, requests a half-million-dollar loan from the United States. Instead the United States grants a million dollars to further Grand Fenwick's space program and show America's sincere desire for international cooperation in space. Not to be outdone, Russia gives one of its outmoded Vostoks. The scientists of Grand Fenwick find that the errant wine crop can fuel this rocket. They send the spacecraft to the moon, beating both the American and Russian teams. The U.S. and USSR spacecraft land shortly after the Duchy's. In hasty attempts to get back first, both Russians and Americans fail to rise from the lunar surface. As a result, Grand Fenwick's Vostok has to rescue both crews.

The British stars, James Moran Sterling and Margaret Rutherford, came to Cocoa Beach for the world premiere, as did Gordon Cooper and his family, and many of the dignitaries of the Cape area. For a moment the tensions at the spaceport ceased, and the men caught up in the space race enjoyed a good laugh at their own expense.

"The first and the most truly heroic phase of the space age ended in the summer of 1963," wrote space analysts Hugo Young, Brian Silcock, and Peter Dunn in *Journey to Tranquility*. "Two years had passed since President Kennedy's commitment to the moon. They were, to the public eye, the years of the astronaut; a period when this strange new breed of man was established as something larger than ordinary life, with gallantry and nerve beyond the common experience."[11] This vision stemmed from the novelty of the situation, the ruggedness of some of the characters among the original seven, and partly, too, from the nature of the Mercury program. "Somehow one man in a capsule, alone in the totally unfamiliar void, more easily acquires heroic status than two or three men facing the ordeal together." The last flight of the Mercury series by Gordon Cooper in May 1963, the authors concluded, "was the last appearance of the astronaut-as-superman."[12]

To most Americans, the astronauts were seen as brave men. The spaceport workers, who realized the risks that the astronauts were taking on a giant firecracker, had a deeper understanding of their courage, as well as their human failings. The astronauts were close to possible danger at all times, and yet the men and women of the Cape saw them push right ahead, just as if it were an everyday job, not too far distant from what the padmen were doing in almost total safety.

That summer marked more than the end of Mercury, as people began to realize for the first time what the moon program really meant. Before that, Kennedy's words had mesmerized them. NASA had gone about its work in an atmosphere of public consent and mute congressional approval. It had decided how to go, where to go, and who should go. The general public accepted the basic lines of the gigantic undertaking. Now qualified people questioned the very concept of Apollo. When the great debate that Kennedy had asked for two years earlier finally got underway, scientists began to see that the space program made distorting demands on skilled manpower, economic resources, and human determination. And they began to ask if it was really worth doing. Was this comparable with pushing the Transcontinental Railroad to the Pacific in the 1860s or cutting a canal through the Isthmus of Panama earlier in the twentieth century? Did we have to beat the Russians? Was this the most important scientific effort we could perform? Was NASA perhaps traveling too fast? The president himself seemed to have his doubts when he began to suggest joint space efforts with the Russians.

Without fanfare, NASA had anticipated the president in this area. In March 1963 Dr. Hugh L. Dryden, Deputy Administrator of NASA, had made an agreement with the Soviets on space communications and meteorology that implied that cooperation was feasible.[13] In an address to the United Nations General Assembly on September 20, 1963, President Kennedy stated that joint U.S.-USSR efforts in space had merit, including "a joint expedition to the moon." He wondered why the two countries should duplicate research construction and expenditures. He did not propose a cooperative program, but the exploration of the possibility.[14] On the next day, Congressman Albert Thomas, chairman of the House Appropriations Subcommittee on Independent Offices, wrote President Kennedy to ask if he had changed his position on the need for a strong U.S. space program. The president replied on September 23 that the nation could cooperate in space only from a position of strength and so needed a strong space program.[15]

Scientists began to talk of other priorities, such as the declining water table in the West and the challenge of oceanography. But geophysicist Lloyd Berkner still took a strong stand for Apollo, chiefly concerning himself with the project as a national motivating force. He had been one of the original promoters of the launching of a satellite during the International Geophysical Year. Berkner's grand vision satisfied many on Capitol Hill. But a majority of scientists still seemed to question the entire program. They felt that the president had proposed the lunar landing in a period of panic that had stemmed from the success of cosmonaut Yuri Gagarin, first man to orbit the earth, and the disaster of the Bay of Pigs just seven days later.

In November 1963, *Fortune* magazine summarized the discussion in an article entitled, "Now It's an Agonizing Reappraisal of the Moon Race."[16] The author, Richard Austin Smith, seconded the president's suggestion to the Soviets for international cooperation instead of the "space race," even though he had not originally advocated it. Smith discussed three levels of attack on the manned lunar landing program. First, a practical view held that the investment of money and talent in Apollo was out of proportion to foreseeable benefits. Warren Weaver, vice president of the Arthur P. Sloan Foundation, had discussed the many alternatives for educational use of the $20 to $40 billion that the moon race was expected to cost. Second, some scientists who were enthusiastic about space exploration feared that Apollo and other man-in-space programs would swallow up the funds that could go to unmanned programs, which they saw as more efficient gatherers of scientific information.

Third, a growing number of scientists had reached the conclusion that no appreciable benefits would come from the Apollo program. Philip Abelson, director of the Carnegie Institution's Geophysical Laboratory and editor of *Science*, the journal of the American Association for the Advancement of Science, had recently conducted an informal survey and found an overwhelming number of scientists, 110 to 3, against the manned lunar project. "I think very little in the way of enduring value is going to come out of putting man on the moon—two or three television spectaculars—and that's that," Abelson stated. "If there is no military value—people admit there isn't—and no scientific value—and no economic return, it will mean we would have put in a lot of engineering talent and research and wound up being the laughingstock of the world."[17]

After quoting Abelson and discussing these three objections to the Apollo program, author Smith admitted that the most persistent justification for seeking the moon was the matter of prestige. He suggested continuing the space program but abandoning the "crash" timetable in favor of one that placed the moon in its perspective as one way-station in the step-by-step development of space. Apollo with a lower priority could provide benefits, while allowing periodic reappraisal.[18]

All during the Apollo days critics complained that activities in space worked against efforts on earth to help the poor and homeless, improve education on all levels, preserve the environment, right race relations, and keep America's reputation as a place of new opportunities for immigrants.

Congressman Olin Teague of Texas, a strong supporter of space exploration in the U.S. House of Representatives, faced this issue obliquely: "'Why should we spend this money to explore space when there is so much to be done here on earth?' Well, there was plenty to be done in Europe when Columbus left it. And there is still plenty to be done there. If Columbus had waited until Europe had no more internal problems, we would still be waiting, but the opening up of the new world did more to revive the European culture and economy than any internal actions could possibly have done."[19]

Dr. Glenn Seaborg, chairman of the Atomic Energy Commission, stated point-blank in a public address of July 24, 1940: "Space exploration, rather than being in opposition to meeting needs on earth, is in fact part of the search for knowledge that is indispensable for meeting those needs."[20] Few could deny that.

The scientific community had valid concerns. First, Apollo was commandeering a disproportionate number of the nation's scientists and engineers to serve its own needs. In his book on the governance of NASA,

Order of Magnitude, historian Frank W. Anderson, Jr., pointed out that administrator James Webb faced both issues openly and effectively.

In recognizing the drain on scientific manpower, Webb won White House support for enlarging the pool of scientists. The government offered thousands of fellowships for graduate studies in space-related disciplines and financed buildings and facilities on university campuses for a new kind of training that the space program needed. During the Apollo years, NASA was to finance the graduate education of 5,000 scientists and engineers at a cost of over $100 million; to spend $32 million for construction of new laboratory facilities on thirty-two campuses; and to make available more than $50 million in multidisciplinary grants to fifty universities. NASA contracts and grants for research were to rise from $21 million in 1962 to $101 million in 1968.[21] This program reduced tensions between NASA and the scientific community, enlarged the research capabilities of the universities, and pointed out avenues of future government assistance to education.[22]

To refute the charge that Apollo served only its own needs and ignored the broader needs of the nation, Webb began NASA's Technological Utilization Program in 1962 to highlight the many items of space technology that proved useful in the civilian sector.[23] A later chapter will discuss these many contributions at length.

3

MOVING AHEAD WITH LBJ

The Torch Passes

On November 16, 1963, President Kennedy visited the Cape for the third time in twenty-one months. He toured Merritt Island and watched a Polaris launching from the *Andrew Jackson*, a nuclear submarine. A week later in Dallas, an assassin gunned down the president. Grief-stricken people quietly walked into their churches on the space coast to pray for their country and its fallen leader, as people did everywhere in the land. They listened to the local radio or stuck to their television sets for details and wonderment at the meaning of the event. Who had killed the president and why? Were other national leaders to be targets? The people spent Sunday afternoon at their televisions watching the funeral cortege move down Pennsylvania Avenue. All remember especially the black horse with the empty saddle and reversed stirrups.

The nation came to call the three Kennedy years the time of Camelot, recalling the site of King Arthur's court in medieval legend. In truth, one had to go back to the days of young Edward the First and his Queen in thirteenth-century England to find such a strikingly beautiful and intelligent young couple as John Fitzgerald and Jacqueline Bouvier Kennedy guiding a people.

The president had said that the torch had passed to a new generation, and he challenged individuals of that generation to look to what they could do for their nation and what they could do, through the Point Four program, the Peace Corps, and Vista, for less materially advanced people. To the men and women of space, he had made a special call: to send a man to the moon and bring him back safely within the decade. And he, the author of that challenge, was not to be there when that moment of triumph came. For surely it must.

Slightly less than a century before, another president, Abraham Lincoln, had fallen to an assassin's bullet. The vice president in 1865, as in 1963, bore the name "Johnson." Both believed in the main concerns of the men they succeeded. But Andrew Johnson in the last century lacked the political backing of powerful men in Congress that Lyndon Johnson had earned by 1963. As Senate leader, Johnson had supported space exploration. Now that he was president, he stated firmly, "We are reaching for the stars. . . . We will not abandon our dream."[1] Two other Texans, both with clout in the House of Representatives, strode forward with him. Albert Thomas of Houston chaired the House Subcommittee on Appropriations and Olin "Tiger" Teague of College Station headed the Subcommittee on Manned Space Flight. Many space-workers looked on the "Tiger" as their strongest supporter during Richard Nixon's presidency. These congressmen and Senator Robert Kerr of Oklahoma had long supported space exploration.

President Johnson gave the name John F. Kennedy Space Center (KSC) to the spaceport under construction for the Apollo flights. With the consent of Florida's governor, Farris Bryant, President Johnson also changed the name Cape Canaveral to Cape Kennedy. The town of Canaveral, however, kept its Spanish colonial name. Several years later, a group of Floridians, led by University of Florida historian Dr. Michael Gannon, worked to restore the historical designation to the Cape. It became, once again, Cape Canaveral, the second-oldest place name in the United States.

Besides his interest in space exploration, President Johnson promoted a wide area of social reform. His term gave great impetus to fairer treatment of minorities in the country by means of fair employment practices and better opportunities in schools. The president's interest in better race relations affected the KSC area. When integration moved into the national orbit, prestigious universities actively recruited all the competent black engineers and scientists. Even one of the better black colleges, Virginia State University, failed to find enough black math and science teachers to staff its own faculty. All members of the math and science faculty were Caucasian or Asian.

Further, black applicants in various categories of work at Kennedy Space Center often lacked the necessary background, training, or union membership for many positions. Thus, while the white community grew rapidly, the black population of Brevard County remained the same, declining noticeably in percentage from one out of four in 1950 to barely one out of ten in 1960.

KSC officials and the contractors tried to improve the situation. They called an Equal Employment Opportunity meeting on April 21, 1964. Most contractors sent representatives. They planned to draw on the local black population, rather than recruit from outside areas. The men at the meeting set up two committees, one for job development and employment, the other for education and youth incentive. This latter area had its special demand: to counter the presumption among young black males that "there was no job for me."

In May 1964, NASA sent representatives to a luncheon meeting sponsored by the contractors' Equal Employment Opportunity Committee. That organization set up a program for employing local black teachers during summer vacation to give them firsthand knowledge of the academic skills necessary for employment at Kennedy Space Center. They could, then, better counsel their students. Principals of three local high schools, a representative of the National Association for the Advancement of Colored People, and a U.S. Air Force Equal Employment Opportunity coordinator attended the meeting.

When Governor Claude Kirk of Florida and his cabinet met with black leaders of the state a year later, Harry W. Smith, the chief of KSC's recruitment and placement branch, took part in that meeting. He explained KSC's Equal Employment Opportunity program. The black leaders commented favorably on the program and hoped that the state government would adopt at least part of it. Here, again, one can see Kennedy's program advancing the state in a significant way.

In spite of the reassurance of the new president, the men and women at the Cape faced a long and difficult year in 1964. Progress on construction of the facilities for the Apollo testing and launching slowed during the long summer months. The fall brought no cheer. On Columbus Day, October 12, 1964, the Russians gained another lap in the space race. Cosmonaut Colonel Vladimir Komarov led a three-man crew in Voshod 1 that included a scientist and a physician and orbited the earth sixteen times. In March of the following year, Americans faced a shock of another kind: an Atlas Centaur exploded on the pad, causing over $2 million in damages. Cosmonaut Aleksei A. Leonov walked in space for five minutes from Voshod 2. The Soviets were increasing their lead. Lunik 4 missed the moon by about 5,280 miles and went into orbit around the sun.

In the meantime, American rockets had improved. The Ranger program started slowly. Early Rangers failed, but Ranger 7, launched on

July 28, 1964, came down to the moon in a sea of clouds. It hit the lunar surface at a speed of over 5,000 miles per hour. But before impact, it had sent back 4,000 pictures. Ranger 8 and Ranger 9 were equally successful. They landed in the Sea of Tranquility and sent back 12,000 photographs. Russia launched its Lunik 9 on January 31, 1966. Less than a minute before impact, its braking system came on, allowing it to land gently in the moon's Ocean of Storms. Within a minute, pictures came back directly from the surface of the moon. The area resembled a volcanic region on earth. One obstacle to future moon landings evaporated. Lunik 9 did not sink into deep dust that some astronomers had presumed covered the moon's surface.

Unions in an Antiunion Area

Contractors had been building houses for the space-workers, structures for the needed services to supply them, stores, schools, churches, roads, and highways all over the area. On the Cape itself they had erected gantries, the launch facilities, and bunkers. All that paled before what lay ahead. They were to build a spaceport, the largest in the world, that would send men to the moon and bring them back safely.

Five factors complicated the relations of labor, management, and government at Kennedy Space Center, especially during the construction years, 1963 to 1965. These were the major issues.

First, most of the nonindustrial states of the South and West had passed so-called "right-to-work" laws. Such laws outlawed both the closed shop and the union shop. The closed shop arrangement required all potential workers to belong to the union. The union shop came about when the workers, under a vote supervised by the National Labor Relations Board, freely chose to form a union. In that scenario, anyone looking for a job had to join the union within a specified time. Many northern industrial states allowed the union shop, but Florida did not. Further, it had gone beyond other "right-to-work" states. It had antiunion provisions in its constitution. As a result, only a public referendum, not a simple legislative action, could change it.

Employers had met "pushovers" when they faced isolated and untrained immigrant orange-pickers or clerks from the checkout counters.

Now they faced tough-minded machinists or electrical workers who had fought labor's battles in Schenectady, Pittsburgh, or Detroit.

Second, while Florida was not an industrial state, the Cape/KSC area, suddenly a large industrial center, found that its labor relations more closely paralleled the practices of industrially developed states. As a result, many labor leaders used the arrangements at KSC as a club over contractors in nearby areas. One of the building trade unions, for instance, jockeyed for advantage with an Orlando contractor by using KSC guidelines as a lever. Should KSC set up noticeably better procedures, workers might prefer to work there, and contractors in other areas of the state would rarely find skilled craftsmen.

Third, many contractors failed to enter serious contract negotiations until workers actually went on a strike. Most of these strikes, fortunately, were short. But contractors could have avoided them altogether had they settled with the union one day before the strike instead of agreeing to union demands after a one-day walkout. Disputes of various kinds held up work at times on the VAB, in the industrial area, and on launch complex 39.

Fourth, jurisdictional disputes caused endless problems. To understand the worker's point of view in this regard, one should remember that the welfare of an entire trade often depended on the protection of certain tasks that came within its jurisdiction. If a trade lost a particular type of work, the union members lost their jobs. Further, union presidents so influenced jurisdictional assignments that unions zealously and carefully protected their existing turfs. Sometimes, these jurisdictional disputes went beyond common sense and outraged everyone. At the same time, management acted on occasion in the "public be damned" spirit of the nineteenth-century robber barons. Industrial accord came with working together.

Fifth, even though belonging to the united AFL-CIO, the construction workers of the old American Federation of Labor, such as the carpenters and plumbers, differed greatly in attitudes from members of the younger Congress of Industrial Organizations, such as the steel, auto, and rubber workers. The more highly centralized industrial unions tended to heed decisions made on a nationwide basis or at national headquarters. The loosely bound construction locals, on the other hand, enjoyed a great deal of autonomy. As one great labor relations man, the Jesuit conciliator Leo Brown, said to his students, "No one can tell the carpenters what to do."[2]

Further, what AFL members got over the years, they themselves won. No one else fought their battles. They concentrated on immediate issues: better pay, shorter hours, and safer working conditions. They opposed company unions as powerless in any crunch. They were suspicious even of profit-sharing plans, since some employers used them as an antiunion device. They avoided politics. America had no labor party as other industrial countries did. Relying on themselves, the AFL unions long hesitated to endorse any political party, while the CIO early formed a Political Action Committee to support candidates.

A large percentage of the American people — especially those in nonindustrial states — viewed unions with suspicion, and many states passed antiunion legislation. Nonetheless, the vast majority of American workers, AFL and CIO members alike, to their undying credit, unlike laboring men in almost every other industrial country, spurned socialism.

The construction workers at the Space Center thought in terms of an immediate job, rather than of the transcendent purpose that lifted the average space-workers above daily routines. Some workers indulged in wildcat strikes, slowdowns, and a deliberate policy of low productivity. They sometimes charged exorbitant overtime costs. International unions failed to discipline the locals.

Contractors, on their part, operating under a cost-plus fixed-fee contract, did nothing to stop skyrocketing costs and excessive overtime payments. The crawler-transporter that took the missile from the assembly building to the launch pad, for instance, cost $3 million more than the original bid of $8 million that had won the contract for an Ohio firm.[3]

Back before President Kennedy announced the moon launch, a permanent Subcommittee on Investigations of the Senate Committee on Government Operations had held hearings in Washington. These led to the establishment by the president of the Missile Sites Labor Commission, with Secretary of Labor Arthur J. Goldberg as chairman. He and three representatives of management were to establish policies and procedures that were intended to improve labor relations within the missile and space industry. It provided for local onsite committees to anticipate problems and prevent them from escalating. The Missile Sites Labor Commission at KSC included one representative of each of the following: the Defense Department; NASA; building contractors; the Building and Construction Trades Department of the AFL-CIO; the industrial contractors; the industrial unions; and the Federal Mediation and Conciliation Service. The work of this committee, coupled with other factors, resulted in a marked decrease in man-days lost at the Cape.

In spite of this, many short-term strikes occurred. For instance, the Brotherhood of Maintenance Workers placed pickets on all entrances to Merritt Island and the Cape at 5:00 a.m. on June 8, 1964, without giving prior notice. Members of the building trades honored the picket lines, closing down all construction work at KSC and the Cape. About 4,000 of 4,500 workers stayed away. Most strikes were purely local, or at most regional, such as one against the Florida East Coast Railroad.

When construction neared conclusion all over Merritt Island, a nationwide strike of Boeing workers hit the Cape. At the Cape, it affected directly only about fifty workers on the Saturn and about 225 on the Air Force Minuteman. W. J. Usery, regional representative of the Machinists Union, tried in vain to prevent the walkout of those machinists who worked for firms other than Boeing. A number of construction workers walked off the job in support of the machinists. All the workers from the Marion Power Shovel Company, who had come south to assemble the crawler-transporter, went home to Ohio. Boeing balked at the union shop request. But the negotiations eventually resulted in a new contract that satisfied the international leadership of the union, and the spaceport machinists voted to end the nineteen-day strike. The skilled conciliator, W. J. Usery, rose steadily in acclaim, and ultimately became Secretary of Labor.

When looking at the work stoppages, one might view those warm months as a summer of discontent. But the work went on, and not at the cost of destroying unions in the name of national purposes, as some urged. Ninety-four percent of the workers on KSC contracts belonged to unions, and they had their hand in building a magnificent, lasting spaceport that continues to amaze everyone.

Alligators, Coots, and Palmettos

A wildlife refuge grew up as a safety belt around the NASA installations. In 1963, the National Aeronautical and Space Administration made an agreement with the Department of Interior's Bureau of Fisheries and Wildlife. Originally an area of 25,300 acres, it took the name of the Merritt Island National Wildlife Refuge. Over the years, the area grew. It eventually included 140,000 acres of land, water, and swamp. The refuge included the entire Kennedy Space Center, except those acres specifically housing spaceport facilities.

A rich variety of wildlife flourished in the area: raccoons, armadillos, bobcats, hogs, and an estimated 2,000 alligators. One of this number, named Casey by design engineers, preferred the lake immediately to the north of headquarters. He generally appeared during the lunch hour to assure his friends that all was well. The large variety of birds attracted many ornithologists as well as amateur birdwatchers. The complete list included 250 species, some of them rare and endangered, such as the bald eagle and the dusky seaside sparrow. Over twenty-two species of ducks wintered at the refuge, with the estimated peak population around 70,000. The wintering coot population exceeded 100,000. Hal O'Connor, and later Robert G. Yoder, managed the refuge.

Between flight and congressional hearings, the *Spaceport News* regularly carried accounts of something unusual in the wildlife area. One year, the men under Hal O'Connor tagged young turtles as they came up out of the ocean onto the beach. The following year, Gordon Harris, Director of Public Relations, urged Charles Johnson, one of the *Spaceport* writers, to watch for the return of the turtles. Johnson had given up a teaching position at Immokalee in the Everglades to become a feature writer on the staff of the *Spaceport News*. He made a careful study of the habits of the turtles. After naturalists verified their tags, the little turtles went back out to sea. Knowing that Harris edited the articles carefully but ignored titles, Johnson called his essay: "Tiny Tagged Turtles Trot Toward Tide." No one recorded Gordon Harris's reaction. But the familiar tongue twister, "Peter Piper Picked a Peck of Pickled Peppers," now had a rival at the Cape.

As the population of the area continued to grow, the automobile remained the only significant means of local transportation. Two roads that figure prominently in KSC plans were the north-south Merritt Island road (U.S. A1A) and the Orsino Road, an east-west street that dead-ended near the Indian River. The industrial area was southeast of the junction of these two roads. KSC improved the Merritt Island road as the main north-south artery within NASA property. A four-lane divided highway, eleven miles in length, extended from a mile south of the industrial area to the Titusville Beach road, about five miles north of the assembly building. Studies by the Joint Community Impact-Coordinating Committee that antedated the Regional Planning Commission had given no indication of the tremendous growth ahead for the residential area on Merritt Island, about ten miles south of the KSC industrial complex. As a result, the State of Florida failed to widen Highway 3, the two-lane road that ran south from the KSC area through Courtenay to the Bennett toll road (Highway 528). After 1964, four FHA-backed apartment complexes spurred extensive residential growth in that area of Merritt Island. As a result, Florida Highway 3 became a bottleneck during peak traffic hours.

The spaceport had not always been open to the public. In 1950, when the military controlled the launch pads, they allowed only those coming on official business to enter. Even the wives of the Secretary of Defense and the Secretary of the Army could not accompany their husbands on visits to the Cape. In 1963, Kennedy's Secretary of Defense, Robert McNamara, authorized the Air Force to let visitors drive through the Cape on Sunday afternoons. In order to give wives and children a better understanding of the activities of their husbands and fathers, Kennedy Space Center's Protocol Office began to hold Saturday tours of Merritt Island and launch complexes 34 and 39 in the summer of 1964. On each of the first two Saturday trips, more than 200 wives and children made the tour.

By late 1964, other visitors besides the families of employees wanted to see the growing wonders of Merritt Island. As a result, on the first Sunday of 1965, KSC began a Sunday tour. Guards handed out brochures and a letter of welcome from Director Debus as the cars passed through the gate. More than 1,900 visitors came the first Sunday, some from as far away as Nebraska and Ontario. As the Sunday tours grew more popular, KSC laid plans for a permanent Visitors Information Center. In late June 1965, a group of architects met with officials to discuss design possibilities, while the National Park Service estimated the potential visitor attendance to be in the millions by 1967. The estimators prophesied well.

Soon the Space Center began to accumulate a long list of famous visitors. At launches, KSC readied a special stand for VIPs. William Cardinal Conway of Ireland testified to the great concern of others for Americans and our activities in space. Present during a Gemini launch, he said that in every cottage in the Irish countryside and in every Georgian home in Dublin, people followed the U.S. space program and prayed for the safe return of the astronauts. "It is our program as well as yours," the Cardinal said. Other distinguished visitors over the years were Premier Georges Pompidou of France, Presidents Lyndon B. Johnson and Richard M. Nixon, Vice Presidents Spiro T. Agnew and Hubert H. Humphrey, Secretary of State Henry Kissinger, His Royal Highness Prince Philip, Duke of Edinburgh, General Joseph Mobutu, the president of the Republic of Congo, Ludwig Ehrhard, chancellor of West Germany, Chief Justice Earl Warren, General Charles A. Lindbergh, and movie stars Roy Rogers and Dale Evans.[4]

To provide a place where all spaceporters could come together on occasion for relaxation, a group of employees developed a recreation area three miles east of Highway 3, halfway between headquarters and the residential area farther south on Merritt Island. Situated on the west bank of the Banana River, with 850 yards of shore line and a boat basin, the tract boasted a setting of live oak, palm, persimmon, and pine trees, and provided playgrounds, picnic areas, and a swimming area. The Spaceport Travel Club also organized a year-round series of trips that specialized in Caribbean cruises and air journeys to Europe, Hawaii, and the Orient. In spite of these efforts, the KSC employees remained compartmentalized, close to their own division or contractor, united only in the purpose of sending men into space and bringing them back.

East-west traffic was never to present a problem. The four-lane divided highway a few blocks north of the industrial area ran east almost a mile, then turned southeastward to a two-lane causeway over the Banana River to the Air Force Missile Test Center industrial area on the Cape; there it connected with the four-lane traffic artery to the Cocoa Beach area and south.

The building of a three and one-half mile long, four-lane causeway across the Indian River to the west connected the Orsino Road with U.S. Highway 1 on the mainland a mile and a half south of Titusville. Originally intended as a limited-access road for KSC-badged personnel only, this road became a public highway a few years later with the opening of the Visitors Information Center a mile and a half west of the KSC industrial area. On the west, beyond U.S. Highway 1 on the mainland, state road builders were ultimately to continue the east-west road as a four-lane divided highway just north of Ti-Co Airport to its junction with Florida State Road 50 near the intersection of Interstate 95. Thus traffic could move rapidly west from the industrial area across the Indian River and on to Titusville to the north; Cocoa, Rockledge, Eau Gallie, and Melbourne to the south; and the suburbs of Orlando to the west.

Brevard Continues to Blossom

The 1960 census recorded 111,435 residents of Brevard County.[5] In May 1963, the Florida Power and Light Company estimated the booming population at 156,688. In the estimates for three years after that, the company expected the 72,650 people in southern Brevard County to admit over 52,000 newcomers; the 54,940 in central Brevard to grow to 100,000; and the 25,760 in northern areas at least to double.[6]

A month later in June 1963, Paul Siebeneichen and his staff at KSC's Community Development Office presented more detailed statistics on the population of the county. By that time, forty-two new residents were arriving every day. Nine out of ten homes were single-family units, and each housed an average of 3.4 people—the statistic the Florida Power and Light Company had used the previous month. The number of men approximated the number of women. Three out of every four men over age fourteen were married. More than one-third of the women over age thirteen had jobs. The median income per family, at $6,123, topped the state. Consistent with this, the median value of homes was $13,000, compared to the state's average of $11,800. While an occasional loner might live out on Apricot Lane or Mulberry Trail, most of the people settled in four towns: Eau Gallie and Melbourne in the south, Cocoa in the center, and Titusville in the north. Even though it was the county seat, Titusville had reached only half the population of each of the other three in the 1960 census.

In 1963, NASA funded three studies of the social and economic development of the area. A regional planning commission looked at roads and water systems; a Florida State University team dealt with community affairs; and a University of Florida research group studied population and economics. The study groups were to finish their reports within two years. The three principal investigators met profitably with NASA's local officials and delegates of NASA headquarters. They further got in touch with representatives of the various Brevard County communities. Florida State University set up an urban research center in the area and published materials developed by the three studies. Newcomers as well as veterans needed their findings.

In May 1964, NASA and the Air Force took a residential survey, by questionnaire, of the more than 28,000 military, civil service, and contractor employees in the area. This study, tabulated by a team from Florida State University, showed that up to that time residents tended to remain where they had located in the late 1950s. South Brevard had 42.1 percent of the population, with 20.8 percent on the mainland and 21.3 percent in the beach areas. Central Brevard had 40.4 percent, with 12.7 percent on the mainland, 15.6 percent on the north beach area, and 12.1 percent on Merritt Island. North Brevard (the general area of Titusville) had 12.4 percent.[7]

The tremendous growth of Brevard County between 1950 and 1960 paled before the growth in the succeeding decade, except in Cocoa, where the growth from 12,294 to 16,110 was a paltry 31 percent. Palm Bay jumped from 2,808 to 6,927—a 146.7 percent increase. Cocoa Beach had a big rise earlier, but now it almost tripled from 3,475 to 9,952—a growth of 186.4 percent. Melbourne grew from 11,982 to 40,236 for a 235.8 percent growth. Titusville jumped from 6,410 to 30,515—a 376 percent leap. But Merritt Island topped them all, percentage wise, going from 3,554 to 29,233—an increase of 722.5 percent.[8] The move of Huntsville's launch vehicle team and design group in the mid-1960s, by itself, brought about 1,000 households from Alabama to Brevard County. Except for forty Boeing families, newly arrived in Alabama, most had lived for some time in the Huntsville area.

Of the foreign-born population in the county, Canadians led the way with 3,296; Germans followed with 3,474; British, 3,224; Italians, 2,279; Irish, 1,193; Polish, 821; Cubans, 543; Swedes, 540; and Czecho-Slovaks, 538.[9] In spite of the best efforts of the Community Impact-Coordinating Committee to provide information about Florida's east coast, relocation proved difficult for many of the newcomers. The families settling in the Titusville area found no large shopping center closer than Orlando. Titusville had only one small department store. The relocation of a large number of people, many of them from urban centers, to the small towns of Florida's eastern coast where newcomers were not always welcome, brought about dramatic changes in the quality of life. A nationally known writer, invited to work on the history project, sold his home in Washington, came down, looked over the area, said: "This is a cultural wasteland," and went back to the Beltway.

The large majority stayed, however, and went to work. According to Table 104 of the 1970 Census, 87 percent of the 9,255 males sixteen and over and 4,043 or 41.9 percent of the 9,644 females of that age level joined in the labor force. The median income for employed individuals in Brevard County in 1960, $6,123, surpassed every county in the state, followed by those with metropolitan areas. Dade County (Miami) was next, with an annual median income of $5,348. Duval (Jacksonville) stood behind Dade. Orange (Orlando) ranked fourth, and Broward (Fort Lauderdale) fifth.[10] Ten years later, with a median income for families of $11,145, Brevard still led. Almost half of the 58,497 families and unrelated individuals in the county—26,568 in all—earned under $10,000 a year, with 8,086 earning between $10,000 and $12,000; 10,074 between $12,000 and $15,000; 13,605 between $15,000 and $25,000; 1,987 between $25,000 and $50,000; and 127 reaching over $50,000.[11]

The building of the Apollo launch facilities proved especially important for the town of Titusville. It stood fifteen miles from the VAB, in contrast to Cocoa Beach, twenty-five miles away. Residents of Merritt Island, south of KSC, that grew so dramatically, could get to their place of work in a short time during low traffic hours. But when they approached in the morning or returned in the evening, the two-lane highway proved inadequate. One slow-moving vehicle could test the patience of the average engineer accustomed to top efficiency.

Titusville drivers met no such traffic congestion. Unless a boat or barge moved under the drawbridge over the Indian River, residents of the south end of Titusville could drive to KSC Headquarters in twelve to fifteen minutes at the current allowable speed. Residents of the north end of town could be at work in half an hour. The proximity of Titusville to KSC brought fast changes. Where no shopping malls stood several years before, two developed six blocks apart on U.S. Highway 1. St. Theresa's Catholic Church stood strategically between the two. Morrison's Cafeteria offered excellent seafood. The Orange Julius stand in the Miracle City Mall lured many thirsty visitors. Soon one could enjoy Kentucky Fried Chicken, Virginia Fried Chicken, or Maryland Fried Chicken within a stretch of ten blocks on U.S. 1 or stop in any restaurant and order Florida fried chicken.

By this time, people had come from all over the nation, not just from the South. Mayor Vern Jansen recalled a meeting of twelve people in Titusville. While all but he would have been Florida-born at such a meeting ten years earlier, now only one claimed to be a native Floridian, and no two came from the same state. An event that brought North Brevardians together was an annual community fair sponsored by the Knights of Columbus on the grounds of St. Theresa's Church. Art Gruenenfelder and his wife Dawn have warm memories of the festivities. The fair brought together all people of Titusville and its vicinity: natives and newcomers, old and young, black and white, Protestant and Catholic, Jew and Gentile. Over the years it grew. Each night, a different nationality sponsored the meal. Cuban night came first, then Mexican night, Italian night, and Chinese night sponsored by the Moon family, who operated beautiful Moon Island Restaurant on the Indian River overlooking the VAB and the launch pads.

While most of Brevard County had relatively few foreign-born residents, Titusville bragged about its cosmopolitan character: Chris (Mrs. Don) Boland was born in the Ukraine, Theodore Poppel in the Rhineland, Emile Stephan in Jerusalem, Hector Figueredo in Havana, Joe and Jeanne Boucher in the Province of Quebec, George Barker in Ulster, and St. Theresa's pastor Father Michael Hanrahan in County Limerick, Ireland.

The big churches in those early days in Titusville were the First Baptist and the First Methodist. St. Titus Episcopal and St. Theresa's Catholic were small at that time. When a group of Auburn University graduates came down from Huntsville, they went briefly to the First Baptist and then started their own church on Park Avenue, soon the largest Baptist church in the area. A number of Methodists came in to form a second Methodist church. A disproportionate number of early space-workers came from the Southern states where the Baptist and Methodist Churches flourished. Since no particular firm came from those cities of the North such as Minneapolis, where numerous Lutherans resided, the two Lutheran churches never had large congregations. A few former St. Louisans belonged to the Missouri Synod. St. Theresa's Parish grew dramatically with the coming of Grumman employees from Long Island, Martin men from Baltimore, and McDonnell families from St. Louis. Its school soon gained a high rating.

The transformation of Brevard County from an essentially rural economy, based upon fishing and citrus growing, to the nation's launch center for lunar voyages and space shuttle flights, occurred principally within the span of twenty years. In 1950, the U.S. census figures revealed 23,653 residents in a sparsely populated county. By 1960, the population had grown to 111,435. As a result principally of the growth of the John F. Kennedy Space Center and NASA, the population soared to 230,000 in the census year of 1970.[12] After man had traveled to the moon and returned, the Apollo program concluded its phase with the ensuing reduction of work forces at the space center.

But while employment declined, Brevard County population rose slightly, so that by 1972 the county had over 240,000 residents, ten times what it recorded in 1950. The needs of space-related employees and their families spurred a tremendous growth in service trades. Their taxes helped to construct schools, city halls, and other county and municipal institutions. New housing areas developed, new shopping centers, hospitals, and libraries, as well as new social and cultural agencies and programs.

In 1958–1972 the federal government spent $2 billion to change the social and economic structure of the area. The 1970 Census figures included more than $52 million to insure housing mortgages, over $13 million in public and elderly housing loans, almost $10 million in grants for public facilities, $50,974,000 for public schools, $74,809 for roads and bridges, $1 million for airport expansion, almost $9 million for hospitals, $1 million for county health centers, $1.5 million for port improvement, and almost $10 million for harbors and waterways. NASA itself contributed directly to some Brevard projects from its appropriations: $2 million for the Cocoa water system, $1,632,000 for the Indian River Bridge at Cocoa, and $700,000 for the Port Canaveral lock. The Department of Defense also contributed to the improvement of roads and bridges needed for the Air Force's Eastern Test Range activities.

Thirty-eight more manufacturing firms opened in Brevard during the 1960s. ABC Canvas made awnings. Americus, Inc. made transformers. Art Kraft specialized in neon signs. Boeing in Titusville produced ejection seats for fighter planes. Other firms provided health care products, medals and awards, kitchen cabinets, signs, filing systems, finger-print scanners, wood trusses, bridge cranes, artificial limbs, plastic automotive parts, sheet metal products, precision tools, garage doors, orthopedic braces, ceramics, eyeglasses, and engines.

Vern Jansen and his wife, Bernice, still traveled to Orlando occasionally, but no longer for basic needs.

Education in Expanding Brevard

In the area of education, by 1950, Brevard County's thirteen schools had an average daily attendance of 4,163. Within five years, that number had doubled. It doubled again in 1960 and again in 1965. By that time, almost 40,000 students learned their American history and English grammar in 1,473 new classrooms, up from the 117 of 1960. The capital outlay for the county schools during those fifteen years rose from $751,800 in 1950 to $26,316,600 in 1965. The 1970 Census listed 75,504 Brevardians between the ages of 3 and 34 enrolled in schools: 1,508 in nursery schools and 3,915 in kindergartens (more of half of each category in private schools); of the 44,721 in elementary schools,

41,329 attended public schools; 17,754 of the 18,246 high school students were in public high schools; and 7,114 attended college.[13]

Religious denominations had opened the first schools in the United States: Congregationalists in New England, Presbyterians, chiefly in the Middle Colonies, Anglicans in the South, and Catholics in the originally French areas in the Mississippi Valley, such as St. Louis, Missouri, and New Orleans, Louisiana. Before the public school movement grew strong in the North, Presbyterians ran most of the religious-based schools. Then, in the 1840s and 1850s, immigrants from Germany and Ireland came into northern cities and the rural areas of the Middle West, and opened parochial schools. Nativists, who denigrated immigrants, denounced parochial schools as foreign intrusions. The Presbyterians, not wishing to face this hostility, gradually closed their schools. Efforts here and there to bring the parochial schools within the growing national drive toward public education for all ran into roadblocks.

The heavy German and Irish immigration in the pre-Civil War period hardly touched Florida. The "Second Immigration" of people from the south and east of Europe went to Detroit and Pittsburgh, not Tallahassee or Jacksonville, much less to Brevard County. Even though Father Thomas Hassett, Irish-born pastor of colonial St. Augustine, opened a school for Minorcan children in the parish in September 1787—apparently the first free school in what is now the United States—the tradition of parochial schools took root in the state of Florida only slowly.

Florida came into the Union in 1845, at a time when the public school movement marched with steady steps under New England and Michigan impetus. While many colleges, especially in the southern states, continued under religious auspices, most elementary and secondary schools were "public," but reflected a Protestant attitude toward life. This was the situation at the start of the space age.

Activities on the space coast changed that. When Grumman sent engineers from Long Island, many of them had family traditions of religious-based education. So it was with people Martin transferred from Baltimore and North American from California. The many families McDonnell transplanted from St. Louis had an even stronger tie to parochial education. In the Missouri metropolis, religious-based education had flourished for over twenty years before public education even got underway in the 1840s. One hundred and fifty years later, parochial and private schools still outranked the public schools in numbers and reputation.

The newcomers from these areas widened the scope of education. Of the seventeen private elementary schools, four were nondenominational, five Protestant, and eight Catholic. Two of the nondenominational schools were in Titusville and two in Melbourne. The Episcopal Parish, Seventh Day Adventists, and the Christian Church sponsored schools in Cocoa. The Lutherans and Baptists opened schools in Merritt Island. Two Catholic parishes in Melbourne and one each in Rockledge, Titusville, Palm Bay, Cocoa Beach, Merritt Island, and Eau Gallie operated parish schools. Melbourne Catholic was the only Catholic high school in the county.

Such an expanse of education facilities challenged the already busy building contractors. "The building construction was slow," Newton Gregg, a graduate of Southern Methodist University, stated in an interview, "but the schools picked up fast because parents were trained and intelligent and the teachers were good."[14] Gregg had worked on dam construction in a remote area of southwest Missouri before joining Donald Buchanan's design group at the Cape. He knew the difficulties of creating a sound educational system in a mobile environment. His son Paul, incidentally, an honor student at St. Theresa's elementary school in Titusville, ranked high in his class at Titusville High School. In the mid-1960s the Titusville High School had so many students that it had to go on split shifts. Those who lived south of Garden Street attended from 7:00 a.m. to noon; those north of Garden Street took classes from 1:00 to 6:00 p.m. Eventually, Titusville opened a second school to the north called Astronaut High. The number of acceptances at the Army, Navy, Air Force, and Coast Guard Academies from Brevard County soon outdistanced that of most school systems in the nation. The list of Brevardians at one of the service academies will come up in a later chapter.

On the level of higher education, the Florida Institute of Technology, a privately endowed institution, opened in Melbourne in 1958. It depended largely on aerospace contractors for support. Within ten years, industry was contributing between $50,000 and $60,000 annually and gave over $100,000 in 1968 and 1970. The institute had 1,700 full-time students on campus and 300 part-time students. Most of the latter worked during the day in aerospace, as did many of its night faculty. The full-time faculty grew from nine in 1964 to 70 in 1971. The adjunct faculty almost always numbered about 100. Of these, three-fourths were involved in aerospace work. The total faculty stood between 100 and 170 faculty members over those years. The enrollment figures for 1969 were slightly under 1,000 part-time students and 1,200 full-time students. A year later, 657 part-time and 1,346 full-time students attended the institute. The pattern of growing full-time enrollment continued in 1971 with only 324 part-time students and 1,759 full-time students.

By that time Brevardians began to say that the entire county was the campus of Brevard Community College (BCC). But when it began in 1960 as Brevard Junior College (BJC), its campus was an abandoned school building adjacent to U.S. 1 in Cocoa. Thirty-one faculty members taught 217 full-time and 517 part-time students. When the Florida State Legislature authorized the college, it also approved Carver Junior College, a segregated school that began in Monroe High School in Cocoa. The "separate but equal" Carver joined with Brevard Junior College in 1963. Beginning in 1965, BJC offered evening classes in Eau Gallie in South Brevard and two years later offered daytime classes in the west wing of Ruth Heneger School in downtown Melbourne. The college authorities planned a permanent Melbourne campus on Wickham Road. By an act of the legislature in 1968, BJC, originally a part of the county school system, went under a district Board of Trustees. Two years later, the legislature allowed junior colleges to change their names. From then on, the school bore the name of Brevard Community College.

In 1972, BCC rented a building on U.S. 1 in downtown Titusville. The Titusville Center offered a variety of adult evening courses. Over the years, BCC established other evening study centers at Cocoa Beach and Patrick Air Force Base. Federal assistance to Brevard Community College was over $700,000 during the years 1967–68 and 1970–71, but considerably less in the two intervening school years.

In 1963, KSC's director, and the commanding general of the Eastern Test Range, had joined in an appeal to the state legislature for additional facilities for higher education in central Florida. As a result, the state chartered Florida Technological University, with its basic plant midway between Orlando and the Space Center. It had a four-year undergraduate curriculum in business administration, education, engineering and technology, humanities and social sciences, and natural sciences. Plans were being made for graduate level studies in all disciplines. The new institution began classes in the fall of 1968, and by 1972 the university had an enrollment exceeding 7,500 students. It continued to grow as the University of Central Florida.

The State of Florida also established a graduate engineering study program called GENESYS, with an off-campus facility located adjacent to Cape Canaveral Air Force Station. GENESYS provided graduate instruction keyed to the aerospace industry's needs at the Space Center as well as in nearby permanent facilities.

Rollins College of Winter Park offered undergraduate and graduate level courses of study in off-campus facilities at Patrick Air Force Base that were available to the community as well as aerospace personnel. Clearly, concern for higher education grew as dramatically as that for elementary and secondary schooling.

4

LEARNING FROM SUCCESS
AND FAILURE

Leaving Politics to the Politicians

A Florida State University survey showed slight participation of new-comers in the political activities of the community or even of the nation. This should surprise no one. The men and women at the Cape, scientists and engineers, concentrated on machines not the machinery of government. During the late 1960s, students at many universities throughout the country protested repeatedly on issues of racial justice and America's foreign policy in Southeast Asia. But the protesters carried their banners and shouted their threats at Berkeley and Cornell, not at MIT or Georgia Tech. Parks College of Aeronautical Technology of Saint Louis University "sailed along calmly at 10,000 feet," while agitators at the mother campus across the river in Missouri threatened to burn down a beautiful mansion on campus because it housed the ROTC office. Frustrated at Saint Louis University, the protesters succeeded in torching the ROTC building at Washington University at the edge of the city.

The administration in Washington favored space exploration. That was enough for the men and women of the spaceport. They should have voted, but like so many of their fellow Americans, they left that privilege to others. Seventy-seven percent of the old-timers and fifty-seven percent of the early arrivals in Brevard County voted in the Nixon-Kennedy presidential race of 1960, but over half the later arrivals passed by the polls without voting. Some had failed to register. Others did not yet meet residency requirements. A few left voting to the non-space people. Newcomers willingly supported non-political community activities, such as the Parent-Teachers' Association, but few sought political control in their respective communities. They wanted the politicians to stay out of the engineering area. They gladly left politics to the politicians, much as the engineering students left the protest marches to the political science majors.

Few space-workers ran for political office, and those who ran and won often found older residents suspicious and uncooperative. Even the visitors at launches who patronized the fast food stores made trouble for the local officials by scattering litter. While thousands gloried in the thrill of successful launches, after the first few spaceshots, Tom Mariani, Mayor of Titusville and a NASA man, remembered chiefly the big clean-up costs that followed.

At community gatherings, newcomers were often surprised to find that veterans of the space coast had not had a chance to get to know each other. One might work on the pad, another in telemetry, and a third in transportation. Well known in their own section, residents of the same area of Cocoa, and worshippers at the same church, they might recognize each other's names, but they had not gotten to know each other personally.

NASA employees gave generously to the United Fund and other charities. Over the years, to counter health and social problems, a considerable number of agencies developed among themselves a Council on Aging, Alcoholics Anonymous, an Association for Advancement of the Blind, a blood bank, a Braille foundation, a cancer society, home nursing services, and mental health associations. Religious groups such as the Salvation Army, Seventh-Day Adventists, and Catholic Social Services sponsored programs. Concerned adults took leadership posts with Scouts programs and the games of the Little Leaguers.

In the area of medical services, in 1965, Jess Parish Hospital in Titusville had two maternity rooms with a total capacity of four beds. Needless to say, some of the soon-to-be mothers occupied beds in the hallways. Surprised that she had a room to herself for her second baby, one mother presumed it was because she had a caesarian section.

Eventually, the four hospitals in Brevard more than matched regional needs. The number of beds available increased from 244 when the Apollo project began, to over 800 by the time man had reached the moon. Both Jess Parish Hospital in Titusville and Wuesthoff in Rockledge moved into expansive buildings adjacent to their original structures in the mid-1960s. The Brevard Hospital in Melbourne opened a completely new facility. With the exception of Cape Canaveral Hospital in Cocoa Beach, all hospitals had special psychiatric beds. With the general prosperity of the area, it might be easy to forget that there were still indigent people. The Brevard County Health Department did not forget these individuals.

Mobility was a main factor in the lives of many. Were they to work at Kennedy on the next stage or return to California, Nebraska, or New

Jersey? Just as an Air Force major might look at Scott Field in Illinois or McCord in Washington as nice places to serve, but not necessarily as the sites of lifetime residence, some looked on the east coast of Florida as a temporary home and had not sold their residences near the Douglas or Boeing central plants. An equal number viewed the area as their permanent home and intended to find enduring employment in the region when their work at KSC ended. Still others lived in constant uncertainty—a factor that influenced their entire family life.

Many McDonnell families faced this uncertainty. Joseph Szofran, an engineer for that firm, described their status: "We didn't get involved with the Apollo immediately after President Kennedy announced the moon shot. We knew that North American had gotten the contract. Everybody was disappointed. . . . A lot of people scattered." Others came.

One of the North American employees who came to KSC in that year, Robert Bielling, had been born and reared on a farm in north central Florida. He had graduated from Lake Butler High School in 1951 and worked for Standard Dredging Company for two years before enlisting in the Air Force at Lackland Air Force Base in San Antonio. While there, he married Cynthia (Cindy) Kenna, a math instructor in the Air Force, the daughter of a Syracuse physician.

Upon completing his stint in the service, Bielling came back to Florida and enrolled at the University in Gainesville. On completing his bachelor of science degree in electrical engineering, he took a position as a test engineer on the Sprint Missile Project with Martin-Marietta Corporation in Orlando. The Biellings' third son, Charles, was born there, to match their three girls. In June 1965, Bielling started work with North American in the Environmental Control and Life Support Systems. His first assignment as Test Engineer with the Ground Support Group at KSC was to set up and approve the functioning of the test equipment. Later, he worked with the Vehicle Test Group to check out the Apollo command and service modules.

All the while, the Bielling family grew, with two more sons born during those Apollo years. As busy as he was at KSC, Bielling found time for his children. He took part in the Boy Scout program as a Webelos leader for Pack 367, supported his daughters in Girl Scouts and softball, and encouraged the boys in Little League baseball and its equivalent, "Pop Warner" football. He and Cindy made sure the children did well at St. Theresa's School.

The Biellings cherished Titusville, but it was not to be their permanent home until Bob had served two tours of duty in California.[1]

Unlike Bielling, who crisscrossed the nation several times in the succeeding years, Jim Voor, a native of Louisville with a degree in Mechanical Engineering from Louisville University, came in 1966 and was to stay. He had worked for ten years on the *Typhon*, a ship-to-air missile, at the Bendix plant in Mishiwaka, Indiana. Work had begun to drop off there in 1965. He then joined Bendix's efforts on the Site Activation of launch complex 39.

Like Bielling, Voor had a large family. Three boys and two girls accompanied Jim and his wife to Titusville early in 1966. He prepared charts and coordinated Bendix efforts in Firing Room 4. Twenty other contractors shared the effort with Bendix. And several other young Voors joined their older brothers and sisters in the ensuing years.

"Operation Big Move"

The men at the Cape celebrated two formal dedications in the spring of 1965. On April 14, the contractor topped out the VAB, previously called the Vehicle Assembly Building, but since February of that year the Vertical Assembly Building. It remained the VAB to the people at the Cape. While Dr. Kurt Debus spoke of it in an almost mythical way, Ben Putney summed up the workers' feelings with the remark, "This is the biggest project we ever worked on. There just ain't nothin' bigger."

The formal opening of the KSC headquarters on May 26, 1965, brought in dignitaries from all over the country. Soon the people who were going to support, maintain, and operate these facilities had begun to move in, along with their equipment. By mid-September "Operation Big Move" had brought 7,000 of KSC's civil service and contracted employees from the scattered sites to Merritt Island, mostly to the industrial area. Forty-five hundred more would move there during the following months, mostly into the VAB. During 1965, the number of civil service personnel at KSC rose from 1,180 to more than 2,500, chiefly through the addition of the Manned Spacecraft Center's Florida Operations under Merritt Preston and the Goddard Space Flight Center's Unmanned Launch Operations Division under Robert Gray.

In 1965, between Gemini flights, the Florida Operations Group of the Manned Spacecraft Center at Houston, that had supervised Mercury launches, moved permanently to what by that time had become the Kennedy Space Center in honor of the assassinated president. Merritt Preston led the group. As Kennedy Space Center's launch operations director, he had supervised the Gemini launches. His team included Walter J. Kapryan, later launch operations director, Paul Donnelly, John J. Williams, and Joseph M. Bobik. In all, 450 civil servants and employees of supporting contractors, especially McDonnell Aircraft Corporation that built the Mercury and Gemini spacecraft, came with Merritt Preston to Florida. His former co-workers, like Joe Szofran, remembered him as a good administrator. Universally liked by all who worked for him, he made decisions promptly and wisely.

A McDonnell employee saw fit to compare Merritt Preston with his own highly regarded boss, John Yardley: "He was equivalent to John Yardley in his capacity. He worked for NASA and Yardley worked for McDonnell. He was a good decision maker, as was John Yardley. He was a very good manager, and I don't think I knew anybody who didn't enjoy working with him." The McDonnell man then told of an experience where Preston and Yardley had approved a new procedure. "Everybody agreed with it," he stated, "because Yardley and Preston had said to do it."[2]

Later in that same year, 1965, another team of missile-men moved to Kennedy Space Center to launch unmanned craft. Ten years before, the Naval Research Laboratory in Virginia had recruited twenty-six civil servants to conduct the Vanguard launches. Absorbed into the Goddard Space Center at Greenbelt, Maryland, in 1965, the highly successful crew, that by then included 120 government personnel and 1,200 employees of supporting contractors, came to Florida later that year. One of the early recruits, Dr. Robert H. Gray, a native of northwestern Pennsylvania, had worked on rocket engine development in Niagara Falls, New York, early in his career. At the Naval Research Laboratory, Gray had occasion to demonstrate outstanding administrative skills.

Back in 1958, on St. Patrick's Day, March 17, the Gray team had launched the Vanguard, the second U.S. earth-observation satellite. After that launch, utilizing rockets called Delta, Atlas, Atlas-Agena, Centaur, and Thor, the Gray team continued its run of success with weather and communications satellites.

In the spring of 1966, McDonnell Aircraft Corporation, the makers of the Mercury and Gemini spacecraft, won a competition with the Redstone Arsenal in Huntsville, Alabama, to develop and produce the medium anti-tank assault weapon, called the Dragon, for the U.S. Army. Upon evaluating the many possible locations to perform this task, McDonnell chose a location that directly affected the space coast in Florida.

Dan Venverloh, a Design Engineer who worked on the Dragon program, recalled that McDonnell had decided in the fall of 1966 to establish the Florida Division of the McDonnell Aircraft Corporation to use in the best possible way the engineers and technicians soon to be released from the Gemini program at the Cape. Further, McDonnell transferred about fifty engineering and manufacturing personnel from St. Louis to Titusville, Florida, to begin an orderly transition of the Dragon program to the Florida Division. Ray D. Hill, Jr., Base Manager for McDonnell at KSC, served as head of the new McDonnell operation in Titusville. R. Wayne Lowe, a McDonnell veteran, project manager for the Dragon, moved to Florida with the team. The Dragon missile program was located at Washington Plaza in Titusville until a new manufacturing plant was ready one mile west of U.S. 1 on the east-west extension of the Orsino road.[3]

Dan Venverloh also recalled that in those early days of the Florida Division, all attempts to find available machine shop facilities in the Titusville area ran into dead ends. One had to travel to Orlando, fifty-five miles to the west, or Palm Beach, two hundred miles to the South. As a result, the men of McDonnell did most of the early machine and processing work in St. Louis until the manufacturing plant opened in Titusville.[4]

All the while, between 1966 and 1968, over on the Cape, America sent up seven soft-landing Surveyor moon probes. Four of these were successful. They sent back over 50,000 pictures that allowed scientists to analyze the moon's crust. Basaltic rock made up the basic surface layer.

That fact came as no surprise. Astronomers had known by that time that the moon had been the site of tremendous volcanic activity in the past. Between August 1966 and August 1967, America sent up five Orbiters. All functioned perfectly. Scientists had a really detailed knowledge of almost the whole of the moon, including the far side. One of these tremendously spectacular pictures of the crater Copernicus won the name "The Picture of the Century."

Outstanding teams, with equally outstanding leaders, Robert Gray and Merritt Preston, served under the leadership of Dr. Kurt Debus of the Missile Firing Laboratory of Huntsville, newly named director of Kennedy Space Center. Eventually, Kurt Debus combined the position of center director with that of launch director. Later, he named Major Rocco Petrone, a West Point football star and a "get-the-job-done" administrator, to the position of launch director for Apollo.

Countless men and women of NASA found their new place of work. Even more significant for many than the physical move was the psychological move from the pads—where they had hands-on participation in the operation—to desks where they directed the actions of others. But the vast majority of veteran space-workers wanted to keep their hands on the job and "in the grease," as they were accustomed to say.

The Gemini, designed by McDonnell, featured a two-man spacecraft designed for flights of long duration with equipment to maneuver and dock with an unmanned Agenda-B, a vehicle that could add power. During the year and a half from March 1965 to November 1966, ten Geminis orbited earth on flights from five hours to fourteen days. On March 26, "Gus" Grissom and John Young cheered the home folks with the first Gemini manned flight of three earth orbits. Three months later, Edward White and James McDivitt went around the earth sixty-six times. During that fantastic trip, White walked in space for twenty-one minutes with a hand-held reaction device. In August of that same year, Gordon Cooper became the first American to man two orbital missions when he went aboard Gemini 5 with Charles Conrad, Jr., for 128 orbits between August 23 and 29, 1965. Seven other Geminis were to go up before the end of the following year (1966). Gemini 6 rendezvoused with Gemini 7 in space, and other Gemini astronauts experimented successfully with docking in space. They proved that humans could function in the weightlessness of space flight and live in space long enough to get to the moon and return.

"Fire in the Cockpit"

From the beginning, men involved in the launches had misgivings about the speed at which the Apollo program advanced. Even more padmen worried about the hatch North American Aviation engineers had designed for Apollo, among them Joseph Szofran, one of the McDonnell team who had worked on Mercury and Gemini. He and his teammates felt that they could have assisted the men of North American Aviation. He stated:

There's no reason to say that the people who took over the program were less qualified. They didn't have the experience themselves, but they had our experience to pull on—reports, records, and such. One thing that they didn't do, which they should have, was to transfer some of the technical knowledge from Mercury and Gemini to Apollo. A case in point: their hatch was not even close to what the Mercury had. It was a slow in-ward opening hatch. It took a full sixty seconds to maneuver. My colleague Sam Benningfield said that he had tried to get North American to change the design, but they absolutely refused to talk about changing it.[5]

James L. Nordby, Jr., the son of a Norwegian lumberman from the State of Washington, had professional reasons to be concerned. An Air Force Physiological Training Specialist in the early 1960s, he had instructed pilots and crew in the hazard of high altitude flight. In 1966, Bendix offered him a position as an Astronaut Rescue Specialist in the large test chambers used for the Apollo Command and Lunar Modules. Jim and the Rescue Team took part in the chamber tests for the Apollo 1 team of "Gus" Grissom, Roger Chaffee, and Edward White. Nordby and his associates felt that they should be at hand for all tests. But NASA officials presumed that problems might occur at higher altitudes, not at sea level.

"We knew of the danger of a high pressure, 100-percent oxygen environment," Jim Nordby wrote. "We also knew the door sealed from the inside pressure. But no one in our organization learned that the launch teams were testing under these conditions with 100 percent pure oxygen. We could neither warn them nor try a rescue."[6]

Among others, Dr. Frank J. Handel, a staff scientist with Apollo space science systems at North American and the author of numerous articles, was worried about the use of pure oxygen in the spacecraft. As early as a year before, General Samuel Phillips, Apollo program director, had misgivings about the performance of North American, the builder of the spacecraft. He had taken a task force to its plant at Downey, California, to go over the programs.

Tom Baron of North American turned in so many negative reports that some of his colleagues jokingly called him "D R" for "discrepancy report." He was a conscientious quality control man, and his work was always good. His boss, John Hansel, knew this and said that Baron was a perfectionist who couldn't bend and allow for deviation. "Anyone knows," Hansel insisted, "that when you are working in a field like this, there is constant change and improvement. The procedures written in an office don't fit when they are actually applied. Baron couldn't understand this."[7] Nonetheless, Baron's reports were heeded by key men who looked into the matter. Baron's accidental death in a train-car accident a short time later complicated rational discussion of the issue.

After many tests in simulated space, the craft looked ready for a test at sea level on January 27, 1967. The astronauts entered the Apollo at 1:00 p.m. Veteran test conductor, Clarence "Skip" Chauvin, ran the procedure for NASA. Soon, astronaut Grissom reported a strange odor in the breathing apparatus and described it as "somewhat like buttermilk." Whatever the cause, the problem proved temporary. Communications, however, were difficult. Several times in the afternoon, Chauvin called a halt. His tour of duty ended at 4:30 p.m.

William A. Schick, Assistant Test Supervisor in the blockhouse at complex 34, reported in at that time. A tall, dark-haired man with the physique of Moses as Charlton Heston played him in the movie, Schick needed Moses' strength for the stretch ahead. He logged events until 5:40 p.m. and then called a halt. By 6:30, the test conductors were ready to pick up the count, when one of the crew, presumably Grissom, moved slightly. Ground instruments showed an unexpected rise in the

oxygen flow into the space suits. Four seconds later, an astronaut, proba-
bly Chaffee, announced casually over the intercom, "Fire. I smell fire."
Two seconds later, astronaut White's voice was more insistent. "Fire in
the cockpit." Even under routine conditions, with no tension, the man
at the door on the left of the cockpit could open the latch only with
great difficulty in about ninety seconds. The astronauts set about carrying out procedures to do that, but their efforts were in vain. Outside, six
spacecraft technicians, Henry H. Rogers of NASA, and Donald O. Bab-
bitt, James D. Gleaves, Jerry W. Hawkins, Steven B. Clemmons, and L.
R. Reece, all of North American Aviation, braved the intense heat and
tried to open the door from the outside with a crowbar. The door did not
budge. Flames drove them back. The astronauts were asphyxiated.

A shock passed through the ranks of the padmen, as if a hurricane
had suddenly whirled in from a clear sky and blue sea, and swept away
everything at the spaceport. Beyond KSC, thousands of space-workers
heard the news on radio, stopped what they were doing, momentari-
ly refused to accept the dreadful news, then as the realization came
that it was true, prayed or cursed, according to their instinctive reaction,
wondered, and hoped. The tragedy shook the spaceport community,
the country, and the world. Would America continue to reach for the
moon?

NASA headquarters chose two men to inspect the burnt-out capsule,
astronaut Frank Borman, who had gone into space on Gemini, and Operations Engineer Ernie Reyes. The inspectors remarked that everyone
had taken safety in ground testing for granted. No one gave any serious
thought to a fire in the spacecraft on earth. They had presumed trouble
in space, but not on the ground. At the congressional hearing, KSC
director, Dr. Kurt Debus clearly accepted responsibility. He admitted
that he and his men had not viewed this type of test as dangerous. He
promised to check their criteria.

With this clear acceptance, Congressman James Fulton of Pennsylvania reassured Debus in these words: "This is why we have confidence
in NASA. We have been with you on many successes. We have been
with you on previous failures, not so tragic. . . . The Air Force had five
consecutive failures and this committee still backed them and said, 'Go
ahead.'"[8]

The House Subcommittee on NASA Oversight, under the chairman-ship of Congressman Olin Teague, held hearings at Kennedy Space Center. The most noteworthy event in the otherwise routine session was the recommendation of Congressman Emilio Dadario of Connecticut that the members commend the technicians who had tried to save the astronauts. They received the National Medal for Exceptional Bravery on October 24 of that year.

The deaths of Grissom, Chaffee, and White reverberated around the world. People paused and prayed. The tragedy made everyone realize how they had taken for granted the bravery of the astronauts and the difficulties of the moon shot. Putting a new twist on Stephen Foster's pre-Civil War song, a journalist headed his editorial: "NASA's in the Cold, Cold Ground."

But NASA was not in its grave. No astronaut withdrew from the pro-gram. No engineer went back to his hometown. President Johnson told the men of Apollo to go on. Americans were still in the space race and on the way to the moon. The men of the spaceport went forward with renewed vigor. North American designed a new hatch, similar to the type on Mercury and Gemini, and handed leadership of its KSC opera-tions over to Tom O'Malley, a Cape veteran since 1958. A native of New Jersey, O'Malley had attended the New Jersey Institute of Technology and worked for Curtis-Wright on turbo-prop aircraft engines. He had come to the Cape with Convair, a division of General Dynamics, and sent up twenty-seven Atlas vehicles, including the one that launched John Glenn into orbit. After the last Gemini flight, he had worked a year for General Dynamics at a Massachusetts shipyard. Then the call came from North American Aviation to take control of its operations at the Cape.

A commanding figure in any group, O'Malley moved with the deter-mination of a Roman centurion. Just as a contemporary in the world of sport, Coach Vince Lombardi, told would-be Packer football players to block and tackle or go home, so O'Malley told his team to do the job or else. When he saw a man failing to follow orders, he threatened to fire him, only to find that the man worked for another firm. After that, O'Mal-ley's men wore distinctive blue shirts with North American Aviation written on the back. He saw that they did their jobs.

At the request of the astronauts, North American Aviation hired competent and colorful Gunter Wendt, closeout pad leader on Mercury and Gemini, to work with Apollo. Gunter spoke with the most unusual accent on the space coast. But his distinctiveness had many elements beyond his speech patterns. Not a Peenemünde veteran, he had served as a tail-gunner with the German Air Force and came to America after the war. He found an aircraft job with McDonnell. When he gained citizenship, McDonnell transferred him to its Space Division. Combining efficiency with persuasiveness, he became closeout pad leader for Mercury and Gemini. His clowning in tense circumstances matched that of rodeo clowns. Just as those quick and daring men saved many riders from the hooves of Brahma bulls, so Gunter's humorous remarks, made doubly funny by his amazing accent, kept the astronauts "loose" in otherwise trying situations. He had become a favorite with them.

"Lucky Seven" and "Ecstatic Eleven"

The first use of the huge complex 39 of the Kennedy Space Center occurred on November 9, 1967. It was a moment to remember as the first giant Saturn V generated the most power ever unleashed by man in a single instance except for the atomic bomb. The damage it did to all the support works also went beyond expectations. Apollo 5 followed on January 26, 1968. A Saturn IB soared aloft carrying the first lunar module, but no humans, aboard. One more unmanned launch, that of Apollo 6 on April 4, 1968, went from complex 39. The spacecraft orbited and returned through the atmosphere.

Astronauts Walter Schirra, Walter Cunningham, and Donn Eisele went up on Apollo 7 on October 11, 1968, spent two hours in orbit, and succeeded in every mission objective. This highly successful flight featured the first live television coverage from a manned vehicle and gave the first view of weightlessness. It also provided one moment of comic relief to the tense people at the Kennedy Space Center. Before closing out at the pad, Gunter Wendt checked the spacecraft. Astronaut Donn Eisele, a good mimic, was seated at the window that resembled a TV screen. Eisele said in broken Gunterese, "Der face on der television screen ist der face of Gunter Wendt." Gunter responded, "Und der *next* face on der television screen that you guys see better be a frogman or youse guys is in trouble." "Frogmen" were the Navy experts who rescued the astronauts when they landed at sea.

The launch went off beautifully. A few moments later, Schirra, the commander, asked Eisele what he saw out the window. Schirra presumed some such remark as astronaut Gordon Cooper had made on his Gemini flight with Charles Conrad: "I see Cuba, and it's beautiful." Instead Eisele asked, "I vonder vere Gunter Wendt?"[9] Naturally, the entire spaceport staff roared with a glee incomprehensible to the general populace.

The Christmas 1968 flight of Apollo 8 lasted 147 hours, traveled 500,000 miles, and electrified the world. Astronauts Frank Borman, James Lovell, and William Anders swept around the moon, took pictures of that dead satellite and even more impressive pictures of our life-giving planet, Earth. They spoke from space. "In the beginning God created the Heaven and the Earth," Anders opened with words from Genesis. "And the Earth was without form and void and darkness was upon the face of the deep. And the spirit of God moved upon the face of the waters, and God said, 'Let there be light.' And God saw the light and that it was good, and God divided the light from the darkness."

Lovell continued, "And God called the light day, and the darkness he called night. And the evening and the morning were the first day. And God said, 'Let there be a firmament in the midst of the waters. And let it divide the waters from the waters.' And God made the firmament, and divided the waters, which were above the firmament. And it was so. And God called the firmament Heaven. And evening and morning were the second day."

Borman read on, "And God said, 'Let the waters under the Heavens be gathered together in one place. And the dry land appear.' And it was so. And God called the dry land Earth. And the gathering together of the waters he called seas. And God saw that it was good." Borman paused, and spoke more personally, "And from the crew of Apollo 8, we close with goodnight, good luck, a Merry Christmas and God bless all of you—all of you on the good Earth."[10] It was a night to remember!

On Apollo 9 on March 3, 1969, David Scott manned the command's service module, while James McDivitt and Russell Schweickart in the lunar module separated in space. The two later came back and docked, thus taking another big step on the way to the moon. Apollo 10 carried astronauts Thomas Stafford, John D. Young, and Eugene Cernan in orbit around the moon. In this flight, as in the previous one, the lunar module separated from the command module and flew within nine miles of the moon, but did not land. In a recollection years later, Cernan described how it felt to be so close to the moon without landing. He was to reach the moon on a later flight.

July 16, 1969, the day set for the launch of Apollo 11, began with an unusually beautiful dawn. A few fleecy clouds scarcely cut the warm sun. The slight wind cheered the assemblage. Gunter Wendt, the close-out man, and a spacecraft test conductor, Clarence Chauvin, checked everything with the help of backup command pilot Fred Haise, Jr. Neil Armstrong entered Apollo at 6:54 a.m. Michael Collins joined him five minutes later in the right couch, and Edwin Aldrin, Jr., climbed into the center seat. The closeout crew shut the side hatch, pressurized the cabin to check for leaks, and readied it. The closeout crew left the pad. Everyone was ready.

The eleventh Apollo stood on the stand, like an eagle on the topmost pine, poised, in control, dominating the country around it, the focal point of a hundred thousand eyes at the Cape, and more on TV. The countdown approached zero. Ignition commenced at 8.9 seconds with a wisp of white smoke, indicating that the first engine had come to life. The earth shook. The great bird seemed to come alive, hovered for a moment, then picked up power as the five engines reached full strength and the hold-down arms slowly fell away. The eagle took off and sped up into the cloudless blue, relentless, unstoppable, to a long-dreamed-of goal. Thousands shouted in joy, knowing deep in their hearts that the flight of the eleventh Apollo was destined to end in glory. The time was 9:30 a.m., the day July 15, 1969. As the moon missile rose slowly and majestically, a voice broke the tension, "The vehicle has cleared the pad." Apollo 11 had gone beyond KSC's control. The men in Firing Room 1 turned for a moment from their consoles to view the spaceship rising over the Atlantic.

Approved to proceed to the moon, the astronauts fired the S4B engine again, increasing their velocity to 23,860 miles per hour. On July

20, Sunday in the United States, Armstrong and Aldrin occupied and powered the lunar module and deployed its landing legs. The two vehicles separated at 1:46 p.m. Kennedy Space Center time, and Collins fired the command module rockets to move about two miles away. Flying feet first, Armstrong and Aldrin fired Eagle's descent engine at 3:08 p.m. Forty minutes later, the command module emerged from behind the moon. Collins reported, "Everything is going just swimmingly."

Even a simulated landing on the moon in the checkout building at Kennedy Space Center proved a memorable experience. How much more did the real approach almost overwhelm Neil Armstrong and Ed Aldrin! When they came close to the moon, they saw that they were heading toward a crater almost the size of a football field. Armstrong took over manual control and steered to a better site. A moment went by, then Armstrong announced to the whole world: "Houston, Tranquility Base here, *Eagle* has landed."[11] The eagle had reached the moon. The year was 1969, within the time frame set by President Kennedy eight years before!

The Apollo program had attained its objective five months and ten days before the end of the decade. No one probed the deepest meaning of these amazing engineering accomplishments as did Anne Morrow Lindbergh, widow of the great aviator, and a penwoman of vision. In *Earthshine*, she spoke of "the new sense of awe and mystery in the face of vast marvels of the solar system, and the feeling of modesty before the laws of the universe that counterbalanced man's pride in his tremendous achievements." Many had remarked that mankind would never again look on the moon in the same way. She thought it more significant that people would never again look at earth in the same way. We would have a new sense of its richness and beauty. She concluded: "Man had to free himself from earth to perceive both its diminutive place in the solar system and its inestimable value as a life-fostering planet."[12]

Just as everyone remembers where he was and what he was doing when he first heard the news of the enemy attack on Pearl Harbor or the assassination of President Kennedy, so everyone can remember where he was as he heard on the radio or watched on the television as Neil Armstrong and Buzz Aldrin landed on the moon. The *Eagle* had landed on the moon, and the men of the *Eagle* and Michael Collins in the command module returned safely to earth.

A plaque in a park at the north end of Titusville carries an unusual tribute to Apollo 11. "Jules Verne's mythical journey to the moon in the 1880s was similar to the Apollo 11 launch on July 16, 1969. Spacecraft dimensions were approximately the same. Both carried three-man crews. The trio were strapped in couches at launch. They blasted off from central Florida, used retro rockets for descent, and came down in the Pacific. Verne's craft was the *Columbird*, Apollo's the *Columbia*. In his role as prophet, Jules Verne had matched the great Isaiah of the ancient scriptures."

Apollo 11 had shown that spacemen could reach the moon and come back. Apollo 12 proved that the success of its predecessor was not a once-in-a-million lucky shot like the perfect game in the World Series. It touched down on the moon near the unmanned Surveyor that had landed three years before. The crew of Charles "Pete" Conrad, Jr., Richard Gordon, Jr., and Alan Bean put in place a complete geophysical station to collect and rebroadcast data from the lunar surface. The crew traveled some distance on the lunar surface and found parts of Surveyor 3. They returned to earth with parts of the Surveyor and seventy-five pounds of samples from the moon's surface. As significant as it was, Apollo 12 came to be called "the forgotten launch," overshadowed as it was by the glory of Apollo 11 that preceded it and the escape from disaster of the launch that followed it.

All Americans learned the story of Apollo 13, even if they did not see the movie of that name. An explosion of an oxygen tank in its service module fouled up the Apollo 13 flight. Because of the rupture, the command module lost its normal supply of electricity, light, and water, and endangered the lives of its three astronauts, James Lovell, "Jack" Swigert, and Fred Haise, Jr. But men of NASA accepted the challenge. Thousands of engineers and technicians across the country, from Downey, California, to MIT in Massachusetts, worked feverishly to solve the problems. The astronauts themselves braved exhaustion and steeled themselves against despair. The repaired spacecraft brought the astronauts safely home. It was the most suspenseful mission in the short history of space flight. All Americans and millions of their fellow human beings prayed for Lovell, Swigert, and Haise, and applauded Gene Kranz at the controls in Houston for bringing the astronauts back to earth. People called it "a successful failure." No astronaut landed on the moon. Nonetheless, the Apollo 13 came to symbolize ingenuity, determination, and courage.

When I asked José González, one of the first men to work with the German scientists at White Sands, what he thought was most significant about Apollo, González remarked: "To get 50 million people to pray that three men might come back alive from space was worth all the money we could ever put into Apollo. We came to realize that people still believed in the provident hand of God."[13]

Apollo 14 landed on a hilly area of the moon. It featured the Lunar Rover, a vehicle designed to move men, equipment, and samples of moon dust across the lunar surface. The cost of the Rover amazed even Senator Everett Dirksen, the Republican senatorial leader, who was to remain in the memory of his fellow Americans especially for his theatrical remark: "A billion here, a billion there, adds up to money." Alan Shepard, the first American in space, commanded that flight, accompanied by Stuart Roosa and Edgar Mitchell. On Apollo 15, David Scott and James B. Irwin used the battery-powered lunar rover to travel across the moon's surface. While Scott and Irwin were riding around in their little vehicle, Alfred Worden was swinging around the moon as module pilot. The Apollo 16 flight of April 17, 1972 sent astronauts John D. Young, Thomas K. Mattingly, and Charles M. Duke into space. Young and Duke traveled across the moon in the roving vehicle.

The flight of Apollo 17 had two special features. First, occurring at night, the launch was indescribably spectacular. Secondly, it carried the first scientist to the moon, Dr. Jack Harrison Schmitt, an experienced geologist, who gathered the greatest number of rocks ever collected on the moon's surface, again making use of the Lunar Rover. Project Apollo ended, a triumph.

"A Broad Range of Benefits"

The restoration of the balance of power in the world outstripped all the lesser, but still important, results of the Apollo program. When Comrade Kruschev carried a chip on his shoulder, President Kennedy's call for the space shot shook that chip. The nation rose to the challenge in an earth-shaking way. America sent men to the moon and brought them back safely. The Soviets had led at the start of the space race, but America had taken the lead on the stretch. It was America's least expensive war, and the most successful!

The success of the Apollo flights brought America a tremendous amount of good will. Shortly after their return from the moon in December 1972, the Apollo 17 astronauts, Dr. Jack Harrison Schmitt, Eugene Cernan, and Ronald Evans, flew on a goodwill tour to lands along the equator. KSC security chief, Charles Buckley, took the opportunity to travel with them. On his return, he spoke with enthusiasm of the crowds at every airport around the world who acclaimed the heroic Americans who had sojourned on the moon.

Other results were not inconsequential.

On December 14, 1971, even before the end of the Apollo project, the Committee on Science and Astronautics of the U.S. House of Representatives had submitted a report entitled, *For the Benefit of All Mankind: A Survey of the Practical Returns from Space Investment*. This report claimed, "The truth is that space research already has produced an extremely broad range of concrete benefits, not only to the American citizenry but also to the people of all nations. The flow of hard benefits has grown from a trickle to a stream, and it is widening to a river as expanding technology uncovers more and more ways of improving man's mode of existence on earth. The product is knowledge, man's greatest asset: knowledge of the universe, of the mechanism at work within it, and of our planet Earth. Aerospace has helped trigger a new renaissance, a revolution of rising expectations."[14]

This report to Congress contains so many memorable quotations from various vantages that a few more properly belong here. Congressman Joseph E. Karth of Minnesota stated in an address to the National Space Club a year before Armstrong and Aldrin landed: "The real importance of the Apollo program is not just the physical act of getting to the moon. Rather, the significance lies in developing the technology to do it. The accompanying advances in our economy, in production of new products, in new factories and new jobs—these are what really matter. Money for space is spent on earth, not in space. The flow of these funds into the economy, and the benefit of increasing knowledge will return manyfold the cost to the taxpayers today."[15]

In an adjacent vein, associate editor Tom Alexander wrote in the July 1969 issue of *Fortune* magazine: "The really significant fallout from the strains, traumas, and endless experimentation of Project Apollo has been of a sociological rather than a technological nature; techniques for directing the massed endeavors of scores of thousands of minds in a close-knit, mutually enhancive combination of government, university and private industry. This is potentially the most powerful tool in man's history."[16]

A writer for the *Christian Science Monitor* for August 29, 1969, wrote, "For the first time in history, great numbers of people realize that man has tools, resources, energy sources, and knowledge to achieve what earlier seemed impossible. . . . This new awareness parallels the experience of Europeans as they expanded to explore and then develop other continents. Europe blossomed in art, science, social and religious reform. Now the Earth as a whole is blossoming in a new awakening and a new reformation that has also many material benefits."[17]

Commentator Haynes Johnson wrote in the *Washington Post*: "Some intimately associated with America's space effort see its greatest achievement as a state of mind. . . . The space program is the clear proof that a nation can set a difficult goal and carry it out. If it has done nothing else, it has demonstrated how America can, when it wants to, marshal its talent, commit its treasure, gain public support, and achieve its task."[18] Haynes Johnson's statement in the *Washington Post* seems so obvious that one might hesitate to comment on it. Yet it was a most significant thing in that so many elements, government, business, private individuals, universities, all worked together for this great goal.

Moving to other tangible benefits, one of the greatest areas of space research was in the field of communications. With international television, all watched the 1968 Olympic games in Mexico and France, and the opening of Expo 70 in Japan. Daily spot news came from all crucial localities in the world: an earthquake on the Pacific rim, a famine in the sub-Sahara. Previously, Americans were aware of what might be happening in Brooklyn; now they could see bloodshed in Bosnia. Domestic communication service by satellite began. In 1970, a communications satellite enabled more than 30,000 doctors in Europe to participate in a three-hour transatlantic conference with doctors in Houston and San Antonio. One might have well called it a closed-circuit convention. The possibilities of advanced satellite communications between medical personnel and centers offered great hope as another way of improving health care.

Within a decade, the communications satellite, TELSAT, proved reliable, flexible, and cost efficient in long-range communication. The Communications Satellite Corporation eventually numbered 114,000 stockholders. As manager of the International Telecommunications Satellite Consortium, it came to share access to the global system with eighty-two other nations. Intelstat satellites bracketed the world from synchronous orbit.

NASA sent its first METSAT or meteorological satellite into orbit in 1960 and continued during the years a program of research towards improving the technique of weather forecasting by space observation. METSAT could save human lives by warning of destructive storms. Before the satellite, a storm could be born unobserved in the tropical seas and sweep into an inhabited coastal zone without warning, driving a wall of water before it and wind to rip towns into kindling. At the turn of the century, a hurricane claimed 5,000 lives in Galveston, Texas. Another killed 4,000 in the West Indies in 1928, and in 1959 a Mexican storm snuffed out 1,500 lives.[19]

The coming of weather satellite systems has brought a sharp reduction in the storm casualties. Take, for instance, Camille, the hurricane that was born in the Caribbean in August 1969, the most intense storm to hit North America in modern times. While she was still in her infancy, space-borne cameras of the Environmental Science Services Administration's (ESSA) satellite system spotted her. Satellites tracked Camille's erratic progress up the Gulf of Mexico until they established her clear path toward Louisiana and Mississippi. Warnings went out early.

Inhabitants had ample time to clear the strike zone. Camille's 200-mile-per-hour winds smashed whole towns and drove property damage into the millions, but killed few. People rightly praised ESSA's satellite system.

The Apollo program called for extraordinary reliability and performance in launch facilities and operations, boosters, propulsion systems, space vehicles, life support systems, navigation and guidance, instrumentation, communications and tracking, and auxiliary support. This technological undertaking reverberated through industry. The technology that led to the invention of an energy-absorbing device used in astronaut couches, led to the development of an O-ring shock absorber that soon went into public use in highway barriers. An infrared camera for testing spacecraft components was used by major tire manufacturers to detect flaws in design. New metals, developed in aerospace research, especially the titanium alloys, are coming into use in oil refineries, where corrosive chemicals soon destroy ordinary steel valves. Aircraft, ship, and automobile manufacturers are interested in an electromagnetic hammer, designed by engineers at the Marshall Center for fabricating the Saturn V and other large rockets.

Not only did space research help existing firms improve, expand, and diversify their products, it also brought about the formation of new companies. It developed products for non-space marketing, after fulfilling NASA's needs with similar technology. The firm, Electrical Capacitators, now realizes two-thirds of its sales in components manufacturing and system engineering in other fields.

Meeting the goals and timetables of the space program demanded an extraordinary management effort on the part of NASA and the aerospace industry. It had to marshal the resources of countless contractors to coordinate their work, to ensure an orderly flow of components to the assembly lines, to build new facilities, to maintain equipment, quality, and reliability, and to perform a thousand other managerial tasks. It had to bring in the participation of the scientific community and scores of universities. This brought about great advances in management techniques. At that time, Kansas City was building a $200 million international airport, using a NASA-developed management system. The Missouri airport construction was being managed by a fully-equipped information center installed on the eighth floor of the City Hall, which closely resembled the facility that Kennedy Space Center used for running the Apollo manned flights and space flight program.[20]

These were the overarching values of Apollo, direct and indirect. A discussion of other fallouts and unexpected results will come up in a later chapter. Other manned space probes lay ahead. But all the while, and cooperating with it, the unmanned launches were moving beyond the moon and planet Mars. Apollo was a step, not a summit.

What was the next big move after Apollo? In March 1972, Dr. von Braun told the Senate Aeronautics and Space Committee that satellites could collect pollution-free electric power from the sun and radio that power to earth. He stated that NASA had studied the idea and found it workable but costly. He said that solar energy cells could be built in space, in an orbit that would keep them over the same spot on the earth's surface.

Sunlight would be converted into electricity by the cells, then beamed to earth as microwave signals and collected on an "antenna farm" on the ground. While costly at first, a time would come when the power scheme became economically attractive. "Maybe by the end of the century," von Braun predicted.[21] Senator Lowell Weicker, Jr., of Connecticut urged NASA administrator James C. Webb to pursue research on von Braun's idea.[22]

A Contrasting National Effort

A comparison of the space program with the other great national effort of the time, the War on Poverty, makes the former seem even grander. Had the strategists planning the War on Poverty been as focused, organized, and informed as their NASA Moon Mission counterparts, who appraised the state of the art before taking initial steps, the country might have won the War.

The strategists should have moved simultaneously on three fronts: to strengthen the family, to establish self-reliance, and to make possible the ownership of productive property. All three called for a change in tactics. Previous help programs were targeted to individuals rather than the family, and promoted dependency, not independence. Such programs as Aid to Dependent Children actually militated against the unity of the family. Other countries, such as Canada, supported the family, not the individual. If direct aid were deemed advisable, it went to the united family trying to "make it."

Just as NASA looked to companies involved in aerospace activities and to universities with strong engineering programs and, in fact, provided scholarships at these schools, the War on Poverty leaders should have called upon agencies of self-help, such as the Credit Union National Association (CUNA), and provided training for hundreds of credit union field workers, instead of the two usually in the field. The poor could have escaped the jaws of loan sharks, learned thrift, and organized their finances. Even the workers at KSC had a credit union.

The leaders could next have assisted the Co-operative League in promoting other co-operative enterprises in both cities and rural areas, as the New Deal had done in spreading electricity to farms in the 1930s and 1940s. Many poor of America could have come to own the merchandise marts they patronized, following the Rochdale Principles of Co-operation.

In those areas where outright relief was indicated, the planners should have asked: "What is being done in their regard, and who is doing it?" After locating the agencies involved in such activities, the leaders should have helped these existing charitable and religious associations accomplish their task, instead of setting up entirely new offices and recruiting and training personnel to staff them.

In earlier housing projects, the planners had never established a core group, a club, a school, a parish, to stablilize new ventures. They deliberately sought a "mix," as if a community were made of isolated individuals with no preexisting relationship. The War on Poverty leaders should have learned from these mistakes. Further, they failed to involve the dwellers-to-be in the building of their homes, although that practice had proved effective in Puerto Rico and on some Indian reservations. They did not look to ownership but to tenancy. As a result, abandoned projects in our cities and towns recall these kindly conceived but ill-begotten ventures. Their sponsors could have learned much by adopting the systematic approach used in the space program.

Why a History Now?

In 1971, even before the end of the moon landings, NASA headquarters realized that the time was appropriate for a contemporary history of Apollo. The people who brought the program to its successful conclusion were still at KSC or at NASA headquarters in Washington. The records were available. Several earlier attempts had gained only partial success. As a result, NASA sent out a call to American universities to provide a two-man team consisting of a senior professor who had published at least one book and an associate with a doctorate in history. As a result, I came on board.

Why did the KSC contract with the University of Florida to handle this history? Up to a point, the local historians who had been at the KSC for years could do a far better job than could newcomers. But they faced four great problems: as members of the Space Center team, they had long since lost contact with the layman's viewpoint; while NASA looked to wide leadership, they would tend to write an in-house chronicle for people at KSC. They'd not enjoy the freedom of action that outsiders could command. Removed as they were from the current mainstream American life, they'd tend to write of Apollo in a historical vacuum, as if it were somehow separate from the days of burning cities, campus riots, and the most unpopular war in America's history. Last, and even more important, the scholarly world beyond the space community would refuse to accept it as an objective study, but would presume it to be NASA self-promotion.

NASA's prerequisite that the senior historian be a published writer added great validity to the project, just as it did when NASA asked the team of Constance McLaughlin Green, a prize-winning author, and Milton Lomask, a teacher of creative writing at the Catholic University of America and at the Georgetown Writers' Conference, to undertake the Vanguard history. A writer visualizes his readers, enters into their minds and their hearts with a sympathetic understanding, and comes to realize what they already know and what they want to learn.

But why write the history then, rather than let time put the facts in perspective? Writing the history of Apollo so close to the conclusion of the program had advantages and disadvantages. The advantages lay in the opportunities for interviews with the participants in the tremendous enterprise, the availability of multitudinous documentation, and the opportunity of weighing interviews with documentation. The disadvantages were those of all contemporary history: the lack of perspective that only time can give; the inability to see ultimate results; and the possibility of causing hurt to some participants. Even if the passage of time might allow for a sounder analysis, the assembling of the story materials at that time was a service for historians of future generations.

A great challenge for us Apollo historians was to write for the average American reader, who was only slightly familiar with science and aeronautics. We ourselves had to learn the meaning of spaceport jargon and translate it for the lay reader. One of the first men we encountered at KSC, Fred Renaud of Titusville, explained that he "brought the bird on the crawler-transporter from the VAB to L-C 39." That was a simple beginning. We heard of command, lunar, and service modules. Yet when we looked up the word module in the 1967 *Random House Dictionary of the American Language*, we found pertinent only its fifth meaning, related to computer terminology: "a readily interchangeable unit containing electronic components, especially one that may be readily plugged in or detached from a computer system."

The workers at KSC also recast the word "interface." Defined in the dictionary as "a surface that lies between two parts of matter or space and forms their common boundary," it grew to include any kind of meeting at KSC. Perhaps the ultimate in space jargon came in the record of a "Saturn V Human Engineering Interstage Interaction Splinter Meeting of the Vehicle Mechanical Design Integration Working Group." In spite of such unusual and time-consuming gatherings, KSC sent men to the moon within the decade and brought them back safely.

At that time, the media had been picturing all astronauts as a combination of George Washington, Robert E. Lee, Francis of Assisi, and the Eagle Scout down the block. We historians interviewed one of the most highly regarded test conductors. He stressed the fact that, like all human beings, astronauts had their various human characteristics: some were democratic and easy of access, some a little bit more aristocratic or of old line military bearing, some professional, some scientific. One veteran of several flights, while neither of commanding physique nor notably handsome visage, made his presence felt in any group. One might wonder why the astronaut made such an impression, but never questioned that he did. The test conductor concluded: "But one of these brave men was a difficult person to deal with, both for the pad men and for his fellow astronauts."

In recognizing this, we historians did not intend to take away from the honor and bravery of that man or cast aspersions on any of this tremendous group of men who went off into space. But we had to tell the story as it was, a story of contemporaries, men who succeeded and men who failed; men who combined brilliance and pettiness; and some of these men were looking over our shoulders as we wrote.

We historians had learned on the university campus to call on the professional services of the librarian and her staff. The KSC librarian, Marian Kotlewski, and her associates helped when asked and occasionally spontaneously offered assistance. The records were there. The participants and witnesses were also at hand and ready to tell the story as they saw it. No doubt a few central decisions had come from secret meetings, but there seemed to be general agreement on all basics. True, the passing of time and subsequent developments might change the view of a particular aspect of the program. Certainly, the numerous flights of the shuttle gave an appraisal of the building of the VAB far different from a view common at the time when launches were infrequent.

Interviews took place at many locales, from Huntsville to Houston, and with all levels of space-workers from broom-pusher to launch director. Some interviews stood out. One such was with historian Dr. James W. Covington, senior member of an earlier Apollo history team; another with Harvey Pierce, president of the American Society of Civil Engineers, a man of commanding presence, who had contributed to the development of the complexes for Mercury and Gemini and looked at the Apollo story from the outside. In all, either individually or as a team, we interviewed 162 persons formally, and others informally at the KSC lunch room, at the post office, the swimming pool, the beach—wherever individuals had remembrances to share. We heard so much acclaim of certain individuals, such as Rocco Petrone, head of launch operations, who had, in the meantime, been assigned to NASA Headquarters in Washington, and Thomas O'Malley of Rockwell, Inc., that a personal interview never seemed necessary. Their leadership in "getting the job done," as well as their methods of administration, were common knowledge.

All interviewees had information, some beyond the ken of others. Recalling his days at White Sands, José González spoke of the relationship of Peenemünde veterans with American GIs that proved "news" to the second in command at KSC. A ski instructor at Winter Park, Colorado, one of the places where astronauts trained, confirmed the opinion of KSC test conductors as to the personality and mannerisms of an Apollo astronaut. Jim Deese, an engineer on early launch complexes, told why he preferred the stationary to the mobile concept. Walter Cooney believed that homesteading on a planet would be as commonplace in 2121 as embarking from Cork for Baltimore was in 1821. Some wondered why NASA was not planning to channel solar energy. The views were as varied as those interviewed. But almost all were frank, clear, and cooperative. We historians overlooked few angles, none deliberately.

On October 4, 1973, in commemorating the fifteenth anniversary of the establishment of NASA, the *Spaceport News* carried a long article entitled "Milestones in Space." It listed every important step in space exploration, from the beginning to that date, except for one item. It omitted the fire of Friday evening, January 27, 1967.[23] Consistent with this omission, the public relations director who oversaw every issue of the *Spaceport News*, wanted us Apollo historians, who had come on board in 1972, to omit the sad story of the deaths of astronauts Grissom, Chaffee, and White.[24]

"History tells the whole story," we insisted. "How can readers appreciate fully the courage of the astronauts if they do not know of the few tragedies that held up the days of glory? How can anyone appreciate the tremendous achievement of the Apollo landings, if they don't hear the whole story, and if they don't realize the dangers those brave men faced in all their tests and flights?"

FIGURE 1. Dr. Wernher von Braun briefs President Kennedy and other dignitaries at the Cape a week before the president's assassination.

FIGURE 2. The first of ten unmanned spacecraft called Mariner explored the vicinity of Venus and Mars.

FIGURE 3. The Intercontinental Ballistic Missile Row helped to keep the Cold War from turning hot.

FIGURE 4. The success of Bumper 8, America's first successful two-stage missile, made people forget the failure of Bumper 7.

FIGURE 5. Many expected disasters in space. The Apollo 1 crew, *left to right*, Virgil Grissom, Edward White, and Roger Chaffee, gave up their lives on earth.

FIGURE 6. The successful launch of Apollo 11 brought Neil Armstrong and Edwin Aldrin to the moon.

FIGURE 7. The crew in the firing room monitor the Apollo 12 liftoff.

FIGURE 8. The successful flight of Viking 1 augured well for exploration of Mars.

FIGURE 9. The men who went up alone—the Mercury astronauts: *left to right, back row*: Alan B. Shepard, Jr., Virgil I. "Gus" Grissom, L. Gordon Cooper; *front row*: Walter Schirra, Jr., Donald K. Slayton, John H. Glenn, Jr., and Scott Carpenter. Six were Mercury astronauts; the seventh man in the photo, Donald K. Slayton, went into space on the Apollo-Soyuz mission.

FIGURE 10. The space shuttle *Atlantis* moves out on the crawlerway towards launch pad 39A.

FIGURE 11. The bountiful boomerang, the fourth flight of the *Challenger*, the tenth space shuttle mission, touched down at KSC whence it launched in February 1984.

FIGURE 12. The shuttle *Endeavor* rides piggyback to Kennedy Space Center.

FIGURE 13. A breathtaking view of the space shuttle *Atlantis* launch from Cape Canaveral.

FIGURE 14. The space shuttle *Columbia* moves out of the Vehicle Assembly Building.

FIGURE 15. This Global Surveyor was designed to study the surface of the
planet Mars.

FIGURE 16. The Vertical Assembly Building, later called the Vehicle Assembly Building, for a time the largest building in the world, goes up on Merritt Island.

FIGURE 17. *Right to left*: President John F. Kennedy, Dr. Wernher von Braun, and Robert C. Seaman, Jr., associate administrator of NASA, watch a Mercury launch.

FIGURE 18. A splashdown, such as that of Gemini 10, called for cooperation among NASA, Navy, and Air Force personnel.

FIGURE 19. The flight of Gemini 5 assured NASA leaders that man had the capability to travel to the moon and return.

FIGURE 20. The *Challenger* crew, shown here in the "White Room," included two women, astronaut Judy Resnik and civilian schoolteacher Christa McAuliffe (*far left*).

5

GETTING A "SECOND WIND"

Lay-offs, Relocations, and Ulcers in Children

Young institutions, like young people, look to their purposes. Old institutions, like old people, look to their processes. Apollo was over, the battle won. Skylab seemed an afterthought. With no immediate goal, the central focus of many Brevardians moved from space to earth, namely to keeping their jobs as layoffs soared toward the ten thousand mark. Army majors pleaded "veteran preference," a ruling originally intended to give returning GIs priority over civilian applicants for available positions in the immediate post–World War II period.

Even before the last of the Saturn launches, the number of personnel at Kennedy Space Center had gone down dramatically, from over 25,000 to 15,000. But the population in Brevard County and surrounding areas continued to rise. A few younger engineers left for Seattle, St. Louis, Long Island, Huntington Beach, or Downey in California, depending on the location of their employers. Hal Roper and Norm Keegan, among others, went back to the company headquarters in California. But more stayed. Some retirees remained in Brevard County. A few retired elsewhere, such as Walt Cooney, who moved to his wife's ancestral Alabama. New industries, such as environmental services, lured the skills of many, among them Major General Dan Callahan.

Veterans who had served on battlefields overseas, at Fort Leonard Wood in Missouri, or at Camp Howze in North Texas, had grown accustomed to moving. Mobility seemed a part of life to servicemen and servicewomen and their families. When ordered, they packed up and left for their new assignments. Engineers for Lockheed or Martin-Marietta expected a change of residence when they completed their contract. In line with this, many McDonnell employees hesitated to sell their homes in St. Louis, and Douglas folks kept their condos in California.

Others viewed Florida as their home and intended to find permanent employment when their work at KSC ended. Still others lived in constant uncertainty—a factor that influenced their entire family life. One could run into KSC alumni in California or Missouri as well as Florida. I had occasion to visit veteran space-workers in California, among them Cindy and Bob Bielling, who had moved to Garden Grove, California, to work on the shuttle program. In the meantime, Bielling's firm, North

American Aviation, had become part of the Rockwell Corporation. Bielling worked in Downey, California, in the Shuttle Orbiter Maintainability group. They appraised the drawings of the vehicle and looked at the design to make sure of ready access for maintenance and repair.

In that plant, Bielling met several of the old Apollo group who had moved to California: Jerome McKenna, who had worked in Tracking on Apollo, Carl Black, Technician Supervisor, Ken Schell, Electrical Engineer in the Ground Support Group, Carl Robb from the Reaction Control Propulsion Group, and a few others. In 1975 the Biellings moved to Lancaster, California, and Bob worked on the building, testing, and checkout of the Orbiter *Enterprise*. After the *Enterprise* "Approach and Landing Test" at Edwards Air Force Base, he worked on the building of *Columbia*, the first Orbiter to go into space.

Bielling's experience merits more attention as it recalls the mobility of so many space-workers. A native of Florida, he was to work at Kennedy during the Apollo flights until 1973 and in California on the early shuttles until 1979, when he was to return to Florida with the *Columbia*. In 1984 he was to go back to Palmdale, California, to work on a Rockwell plane, the B1 Bomber, for the Air Force. In 1989 he returned to KSC with the Shuttle Logistic Depot that Rockwell was developing for NASA. He was to retire in Titusville in March 1992. While this is moving ahead of the story of space exploration, the narration gives us a picture of one family's contribution to space successes.[1]

The uncertainty of place of work in the near future, combined with the tensions inherent in space exploration, tended to affect family life in many ways. Brevard's divorce rate soared along with the Apollos. Articles in the local newspapers and national magazines regularly carried features on the domestic strain in the space communities. As *Time* magazine was to state: "The technicians who assemble and service the rockets have chosen a tense career, and it has taken its toll on their personalities, their marriages and their community. . . . The rhythms of life at Cape Kennedy are set not so much by the clock or the seasons as by the irregular flights of the missiles. Bouts of furious activity and 14-hour days may be followed by periods of idleness."[2]

The *Time* article saw some difficulties stemming directly from the nature, training, and background of people in the engineering profession. Many engineers were perfectionist males, surrounded all day by scientific precision, who could not brook the sight of an unwashed coffee cup in the sink on their return home. Many carried their work home with them, spending the evening hours not with their families but in reading technical material. Intelligent, but not liberally educated, their interests focused primarily on the technical world.

On reading this article, one housewife remarked: "My husband lines the wall of his den with NASA plaques and citations, but he has yet to wash a coffee cup." Another stated: "My brother-in-law had a big part in putting Apollo together, but it took him two months to put the screen door back in order." And still another: "My ever-loving spouse made the change from Gemini to Apollo, but he can't change our third baby's diapers." "Our boys will be beyond their scouting years," a concerned mother stated, "before their father has time to take them camping."

In an interview, Director Debus spoke frankly of this issue: "There is so much tension, so much anxiety in putting men into space. Yes, we've lost men because of family problems. When a man is so dedicated that the NASA program becomes his personal life, it takes much time away from wife and children. We need a great many understanding wives here . . . in the end we usually have to tell them their husbands will be working even harder next year. Such exposure to stress is rare elsewhere. We live with it constantly. In fact, it is so much with us that we are studying it—how it is affecting our hearts, our nerves, our functions, our aging processes. We don't know yet."[3]

One space-worker said with a grin: "When we're fed up, we can't take the family to Florida to relax in the sun. We're already there."

Putting men into space caused grave family problems. But readjusting to the decline in employment that followed was to cause even greater problems, especially to children. A prominent pediatrician of the region, Dr. Ronald C. Erbs of Titusville, noted a high incidence of ulcers in children, especially during the last half of the Apollo program, when uncertainties arose about their future. Where would they live? They'd miss their friends. "Before coming to this area," he stated, "I did not see ulcers in children, except for rare examples."

Then, on reflection, he added: "It is my opinion that the life generated by the space program was basically unhealthy for the families of space personnel. . . . With the decline of the space program, these highly trained men became very insecure regarding their futures. It is extremely difficult to keep the emotions of work away from the emotions of the family, hence increased family tensions. These tensions then were felt by the children, and since the problems were not usually discussed, the children had no outlet for these emotions, leading to the development of ulcers."[4]

Even earlier in the Apollo years, Dr. Margaret Moore, M.D., of Titusville, spoke of a similar pattern. In one particular subdivision, five or six children had ulcers. As the physician explained, the parents were extremely high achievers; consequently, their pressures were passed down to the children. Ida Reyes, wife of Ernie Reyes, explained that one of their children woke up in the morning and complained of a "tummy ache." After several office visits, the doctor determined that it was, as with several other youngsters in their new neighborhood, a case of ulcers.

Compensating factors were many. First, few people anywhere in the world shared the opportunities of taking part in such a tremendous enterprise. Second, most individuals moved into the area from distant locations, making a fresh start. Third, few "old settler" groups snubbed the newcomers. Most Brevardians welcomed them. Last, everyone had a task, and each task had importance. Hal Roper at his computer in the launch control center and the man who closed the hatch had important jobs to do for the success of the program. This attitude carried over to community life beyond KSC, to the PTAs, to the Baptist assemblies, and to meetings of the Knights of Columbus, Rotary, and Elks.

Ron Nazaro, an employee of IBM at KSC put it well when he asked, "Where else in America would my closest friends be two men who make twice as much money as I do?"[5] One of them was deputy director of security for all areas of KSC.

Business Booms in Brevard

Apollo had changed the face of Brevard County. In 1950, of the 2,771,305 individuals who called Florida home, 23,653 resided in Brevard County, slightly under 1 percent of the state's total population. By

1960, when an increase of 371 percent had brought Brevard's population to 111,435, the state's total had soared to almost 5 million. During the next decade, Brevard did not match that percentage growth, but still led the counties of the nation. By 1970, when the state had 6,673,098 residents, an increase in Brevard of slightly over 100 percent doubled the population to 230,000. Of every 100 Floridians, three then lived in Brevard.[6]

Looking back again to 1950, an observer finds only 452 Brevardians employed by twenty-six small manufacturing firms. All but three had less than sixteen employees. Two had twenty to twenty-nine, and only one, the Harris Corporation plant in Melbourne, had over 100 workers. Four hundred and four industrial workers received an average yearly salary of $2,222. The value of manufacturing in Brevard was $2 million. The capital expenditures were $36,000. The payroll totaled slightly over $1 million. Where did this put Brevard County in the state ranking? Thirty counties outranked it, two were tied with it, and thirty-five counties lay behind. Thus, Brevard placed somewhere in the middle.[7]

As Apollo grew to manhood, the number of industries in Brevard County grew from twenty-six in 1950 to 107 in 1960, and to 215 in 1970. The number of industrial workers grew from slightly over 400 in 1950 to 7,300 by 1960 and to 17,300 in 1970.[8] The work force in Brevard began to look like that of Pittsburgh in the day of Andrew Carnegie. In a state that had 21,303 retailers, Brevard had only 536 in 1950. Forty-eight of these employed hired help. Owners ran the others. The annual sales amounted to slightly under $21 million. Eighty-seven sold food, eighteen general merchandise, thirty-two clothing, twenty-eight furniture, 180 pharmaceuticals, and seventy-five other needed goods. Brevardians could dine at eighty-four different restaurants or drink at any of thirty-eight pubs. By the census of 1972, the number of retail outlets rose to slightly over 2,000, including 341 service stations, 301 restaurants, and 264 food stores. The same census showed 1,762 other businesses offering various personal services: 455 barber shops, shoe repair shops, and so on, 359 business services, 189 hotels and motels, 207 auto repair shops, and 169 automotive dealerships. The wholesalers at that time included forty-six establishments doing a yearly business of $1,300,000 with an average payroll of $27,753.[9]

By that time, Jim Finn, who, when he came with the pioneers, had to drive many miles to get a good haircut, now had his choice of barbers not far from home. Agnes McLearn in Cocoa Beach and Emily Sasko in Titusville could get their groceries within walking distance of their residences. Charley Wingertsahn made sure electrical power was plentiful in the many new residences, and Donald Boland looked for a second career.

Boland had come to Brevard County from Miami in 1957 to build homes. By 1976, his son David, boasting a civil engineering degree from the University of Florida and pointing to four years' experience with the firm, was ready to take over the headship of Boland Construction. Under David's leadership, this once local firm began to undertake work in other areas of Florida and then to bid on construction in neighboring states. In a few years Boland built buildings in eleven states, the most distant being Oklahoma. Many Brevard firms were offering their expertise and services to other areas.

The service trades employed 1,276 individuals in 1950, 10,450 in 1960, and matched manufacturing with 17,300 in 1970. Two thousand individuals worked in retail and wholesale trades in 1950, 6,350 in 1960, and 16,100 in 1970. The number of government workers likewise grew dramatically, from 550 in 1950, to 5,800 by 1960, and to 13,800 in 1970.

As the total number of households in Brevard rose from 7,553 in 1950 to 32,655 in 1960, and to 76,600 in 1970, the most noticeable change came in the percentage of oldsters. In 1950 over one out of ten residents of those households was over sixty-five. By 1970, only three Brevardians out of 100 were seniors. Brevard had become a young community.

The First Research Corporation drew up a table on the impact of Project Apollo on employment and housing in the state. The statistics on Brevard surprised no one. But even some Brevardians read the report on other counties with surprise. Orange County, west of Brevard, including the city of Orlando, gained 15,150 space-related wage earners and 14,000 new homes. Volusia, to the north, including the city of Daytona Beach, gained 13,200 space-workers and 11,000 housing units.

Apollo's influence reached beyond the immediate area of the east central section of the state. Dade County almost matched Brevard with an increase of 31,250 employees in space-related occupations and 28,000 new residences of space-workers. Broward and Palm Beach Counties also ranked high and gave a total for the southeast section of the state of 56,900 space-workers and slightly over 50,000 new housing units. West central Florida, especially Hillsborough County, also felt the influence of Apollo with 27,700 employed in space-related activities and 25,250 new space-related residences. Of 3 million employed Floridians, 200,000 had a hand in the moon shot. Of the slightly less than 3 million residences in the state, 180,000 housed space-workers.[10]

Apollo promoted industry far beyond the borders of Florida. Besides the major contractors on the Saturn project, many subcontractors had their part. In *Stages to Saturn*, Roger E. Bilstein carries an appendix listing the major subcontractors. Two hundred and forty subcontractors had their share in building the booster, and over half of them (132) were located in California. Boeing sent out contracts to fifty-two firms in sixteen states. Douglas sublet work to ninety-one firms in eighteen states. IBM worked with fifty-six firms in fifteen states, and North American dealt with ninety-one subcontractors in twelve states. Only twenty states hosted no space suppliers or auxiliaries.[11]

In 1973, Brevard adopted a county seal with a moon landing scene that represents the technological wealth created by the presence of NASA's Kennedy Space Center and the Air Force's Eastern Test Range.

Skylab and Apollo Soyuz

Back in 1965, NASA had set up a post that eventually became the Skylab Program Office. During the next four years, that office worked out the form of spacecraft and the content of the program. As long as the Apollo goal remained underway, Skylab was a stepchild of manned space flight. But when it became clear that America's space program had to lessen the urgency of Apollo, Skylab became the bridge to sustaining manned space flight while the nation opened up new avenues to space exploration.

Back in the late fifties and early sixties, before President Kennedy challenged the country to go to the moon in the decade, the von Braun team at Marshall Space Center at Huntsville, Alabama, had given attention to putting up a space station in earth orbit. Various stages of the moon shot could be fueled from this space station before moving on to the moon. The president's time frame of a decade ruled out the slow pace that this type of program entailed. When the Apollo program neared its end in 1972, the men and women at Kennedy Space Center turned again to assembling and launching a space station dubbed "Skylab." The Russians had successfully launched their Salyut space station in combination with manned Soyuz launches in 1971.

On July 19, 1972, the first segment of the Skylab arrived aboard NASA's large transport plane, the "Super Guppy," and moved into the operations and checkout building. In close order, all of the components that made up the spacecraft arrived on schedule. The preassembled airlock module/multiple docking adapter, the fixed airlock shroud, and the deployment assembly for the Apollo telescope mount came from St. Louis, Missouri; the orbital workshop and the payload shroud from Huntington Beach, California; the command service module from Downey, California; and the Apollo telescope mount from Huntsville, Alabama. The booster for Skylab reached KSC from New Orleans, Louisiana, aboard the barge *Orion* on July 26 at the time of the Apollo 17 rollout. In September of that year, engineers started stacking the Skylab components onto the Saturn V in high bay 2 of the VAB. The huge Saturn V was to put the space station into orbit. Then a Saturn IB, a much smaller vehicle, was to carry the astronauts in the command service module to work in the orbiting Skylab.

To launch the smaller Saturn IB from launch complex 39 designed for a Saturn V, engineers proposed a 130-foot metal framework so that the second stage, the instrument unit, and the spacecraft stood at the same height as the larger Saturn. After considerable discussion, engineers designed a 250-ton pedestal that became Skylab's most distinctive feature. Four legs of steel pipe more than two feet in diameter supported the launch table. The Saturn IB stood on top. The people at KSC called the pedestal "the milk stool." Paul Bunyan might have used such a seat had he been a dairyman instead of a lumberjack. It looked like a supersized erector set.

KSC intended to give the contract to the major firms in building the pedestal. At this juncture, however, the Small Business Administration in Washington asked that the contract be set aside for one of the small firms. KSC refused, stating that the project needed an experienced total organization to prevent delays. Discussion took up a month. Eventually, headquarters in Washington ruled in favor of the KSC's position. When the bids were opened, to the surprise of many, a small electric firm in Titusville, the Holloway Corporation, had submitted the low bid. Further, the winning proposal called for fabrication by another small firm, this one in Jacksonville. Holloway and its associates finished the work on time and according to specifications.[12]

Brevard County had won a victory. A small firm in Titusville, where no industrial firms existed twenty years before, was now able to match the "biggies" from all over the nation. This victory pointed to the changing industrial pattern on the space coast.

The launch of the Skylab on Saturn V came on May 14, 1973. One minute into the flight an accident occurred that disabled one of the solar panels. Flight controllers maneuvered the spacecraft to minimize damage from excessive heat. Their ingenuity and perseverance saved the $2.5 billion program. In ten minutes, Skylab was in orbit, and in ninety-three minutes it had circled the earth. Eleven days later three Navy men, Captain Charles Conrad, Jr., as commander, Commander Joseph P. Kerwin as scientist-pilot, and Commander Paul Weitz as pilot went up on Skylab II. They faced plenty of trouble in space, but were able to work things out during their twenty-eight-day sojourn. A month later, three other astronauts, Apollo veteran and Navy Captain Alan Bean, Dr. Owen K. Garriott, and Marine Major Jack Lousma, worked in the space station and occasionally took space walks during almost two months aloft. The last Skylab mission featured Marine Lieutenant Colonel Gerald P. Carr, scientist-pilot Dr. Edward G. Gibson, and Air Force Lieutenant Colonel William R. Pogue.

At that time, March 1973, the comet discovered by and named for a Czech astronomer, Lubos Kohoutek, caught public attention. While it did not create the expected thrill for earthbound viewers at its arrival around Christmastime, it did have dramatic visibility from the Skylab.

Three astronaut crews, nine men in all, lived and worked in Skylab for a total of 121 days between May 25, 1973 and February 8, 1974. They conducted nearly three hundred scientific, medical, and engineering experiments and overcame a host of problems—illness, bland diet, extensive work schedules prepared by mission planners, and getting accustomed to weightlessness. Some scientists urged the launch of the second Skylab, even if it did no more than continue the work of the first. NASA, facing a period of shrinking space budgets, provided funding only for the meeting in space of an American Apollo and a Soviet Soyuz. The second Skylab was stored at the Marshall Space Center in Huntsville, Alabama, with no budget to launch it. Subsequently, it was moved to the Smithsonian Institution in Washington, D.C.

In spite of the problems of its early development, Skylab held to its primary purpose of putting man into orbit to perform scientific work, and in that aim it succeeded. Skylab clearly showed that it was feasible to live extended periods in orbit without becoming disoriented or encountering major problems with the lack of a gravity field. Motion sickness hit five of the nine Skylab astronauts, but only in the early stages of flight. The crew exercised on bicycles as energetically as they did on earth. According to Skylab historians, David Compton and Charles Benson, "The longer flights showed that after the first 30 to 50 days, astronauts generally built up a tolerance to the in-flight testing."[13]

Some tasks proved easier without gravity. Moving massive objects, for instance, was no problem as long as there were adequate handholds to control them. Small objects were more troublesome. Hand tools, screws, and other small parts didn't stay put. Often air currents in the workshop carried small objects to the screen covering the intake of the ventilating system. Among the positive results, the solar observations of the Skylab crew delighted the members of the American Astronomical Society, as did the photographs of solar flares. The astronomers dubbed one the "Gibson Flare" in honor of Dr. Edward G. Gibson, who was a member of the third crew with Carr and Pogue.

President Kennedy had talked of cooperation with the Soviets in space during his last visit to the spaceport in the early fall of 1963, a short time before his assassination. This talk of working with the enemy disturbed many powerful individuals connected with space exploration, but no one openly saw a connection between the hoped-for cooperation and the assassin's bullet.

President Richard Nixon had discussed the possibility with Premier Aleksey Kosygin of an Apollo rendezvous and docking with a two-man Soyuz. The differing docking mechanism and life support systems caused problems. But when problems are recognized, they can be solved.

The last Saturn went up on July 15, 1975, with astronauts Thomas P. Stafford, Vance Brand, and Donald K. Slayton aboard. Several hours earlier, cosmonauts Aleksey Leonov and Valery Kubasov had lifted off from the Soviet cosmodrome, the Cape's counterpart in Central Asia. Forty-five hours later, Russian and American vehicles docked in space, presaging the fulfillment of President Kennedy's hope for international cooperation in space and pointing to the end of the Cold War that came a few years later. The two teams welcomed each other in Russian and in English, exchanged gifts, spent two days in scientific experiments, and returned safely to earth.

Commenting on this momentous event, space historian Roger E. Bilstein wrote: "Partnership in space, by itself, will be no automatic guarantee of international amity. Partnership in space exploration may be an exhilarating prospect, however, offering an additional incentive for international cooperation and peace."[14]

Three years after the last Skylab launch, the vehicle again became the focus of media attention. At that time, an unmanned Soviet satellite exploded over northern Canada, scattering pieces of its nuclear-fueled electrical power module over a wide area. The men of NASA rightly began to worry. What might happen if Skylab came down over Chicago's north shore, Kensington Park in London, or the center of Calcutta? Could the men and women of space control their dying spaceship?

Nine years after the last human visit to Skylab, the vehicle began to drop lower and lower toward earth. The time of reentry approached. Naturally, people were worried. Many expected it to fall over the ocean. It passed south of Africa. Finally, sections fell southeast of Perth, Australia. Officials waited anxiously for news of injury or property damage.

No bad news came. Skylab had landed without harm to man or woman, cow or kangaroo, ranch house or cattle barn.

Meanwhile, just three days after Skylab's reentry, two Soviet cosmonauts aboard Salyut 6 established a new record for endurance in earth orbit. The record they broke was not Skylab's, but one that had been set only the year before by another Soviet crew.[15] Between April 1971 and 1985, the Russians launched a total of eight Salyut stations and forty-three manned missions. Thirty-five cosmonauts lived and worked aboard the Salyuts. Svetlana Savitskaya, the second Soviet woman to venture into space and the first to make two flights, was the first woman to walk in space. Cosmonauts from ten countries spent time on the Salyuts as guests of the Russians.

No "Spring on Mars"

While the manned space explorations galvanized the attention of the general public, NASA's unmanned space program compiled its own record of achievement. During a period of thirty years, NASA conducted over 300 launches for programs ranging from exploring the solar system to improving weather forecasting, advancing global communications, and studying earth resources. These projects more than repaid the nation's investment in time, money, and technical talent.

Besides expanding its fleet of unmanned rockets, NASA built powerful boosters capable of performing spectacular feats. Surveyors had landed softly on the moon and helped pave the way for later Apollo missions. The Mariners provided detailed photographs of the cloud tops of Venus and the surfaces of Mars and Mercury. Two Viking landers descended to the surface of Mars and searched for evidence of life, while two Orbiters mapped almost the entire planet from overhead. Two Pioneers went to Jupiter, and one flew on to Saturn. Two much larger Voyager spacecraft followed them to both planets. One of these visited Uranus and Neptune as well. NASA embarked on an ambitious plan to send nine spacecraft to Mars over a decade on Delta rockets and a probe to Saturn on a Titan Centaur launch vehicle. It was a thrilling record over the years. All the while, Russia shared the probing and exploring.

By far, the largest number of unmanned spacecraft remained in orbit around the earth. Communications, weather, and other types of satellites crowded into geosynchronous orbit, that region above the equator, at about 22,240 miles altitude, where a velocity of 6,878 miles an hour towards the east will keep a satellite apparently motionless in the sky. Weather and other earth-observation satellites patrol the planet steadily in a north-south direction on polar orbits. Scientific explorers in many types of orbits have returned a wealth of information not attainable in any other way.

Unmanned exploration of the solar system proved to be one of the most exciting and scientifically rewarding aspects of the nation's space program. The unmanned launch operations continued to place the large majority of spacecraft in orbit until the Space Shuttle became operational. The first commercial communications satellite, launched for the American Telephone and Telegraph Company, had gone up in 1962. The first international communications satellite, Intelsat, went up on April 6, 1965. It was the largest international carrier in history. Pioneer 10 in 1972 flew by Jupiter, taking photographs of the surface and measuring radiation emissions. In that same year, 1972, Landsat 1 pioneered in assessing earth resources from outer space. Pioneer 11 flew by Jupiter and Saturn. Mariner went by Venus, then pushed on to Mercury and completed the first three flybys of that planet.

Up to thirty years ago, men believed that Mars might have some life on it, not necessarily intelligent beings, but trees, flowers, and other evidences of vegetative life. The words of a popular song told of spring on "Jupiter or Mars." While the Russians had little success with flights to Mars, American probes succeeded. Mariner 3 of November 1964 had gone out of control almost at once, but its successor, Mariner 4, on November 28, 1964, stayed on course. By July 14, 1965, it was within 6,000 miles of Mars. Before going into a permanent solar orbit, it had returned twenty-one pictures as well as an amazing amount of miscellaneous information. The best part of the Martian program came in 1971, when after the failure of Mariner 8, Mariner 9, launched on May 30 of that year, went into orbit around Mars at about a distance of 850 miles from the surface.

It was the first spacecraft to orbit another planet. It sent back photographs that showed no signs of vanished civilizations, no pictures of airstrips whence "Men of Mars" could send hostile or even friendly expeditions to earth. One of the first features to appear was an enormous crater eighteen miles deep and wider in diameter than the state of Georgia. Huge canyons stretched across an area equal to the distance between St. Augustine, Florida, and San Diego, California. No signs of life appeared, but interest in Mars grew.

The Viking program of 1976 was the most ambitious of all interplanetary missions up to that time. Each Viking consisted of an Orbiter and a lander. Two Vikings went up in the late summer of 1975 with the landers riding piggyback. They flew to Mars and went into orbit. A year later, both landers were sending to earth information from the surface of Mars. A photograph taken on July 23, 1976, showed sand dunes and large rocks, but no tulips or palm trees. At one and the same time, the Orbiters were mapping the planet and relaying instructions between the scientists on earth and the landers on Mars. While the Viking project found no life on the planet, it showed the great possibilities of exploring the universe with robot-spacecraft.[16]

Back in February of 1961, the Russians had taken the lead in the exploration of Venus with a successful launch of Venera 1. Trackers lost contact after a few weeks when the probe was 4,650,000 miles from earth. They never regained contact. Though Venus is the nearest planet to earth, it is always at least 100 times as far away as the moon. This alone makes contact much more difficult. To confuse matters even more, probes cannot utilize the shortest route and require a complicated procedure that lies beyond the scope of this book. Obviously, a journey to Venus would not be a weekend excursion.

The first American attempt, Mariner 1, in July 1962, had plunged into the sea. The second, Mariner 2, that went up on August 26, 1962, had bypassed Venus at 21,750 miles on December 14. It sent back invaluable data before it moved out of range and entered a permanent orbit around the sun. Trackers lost contact with it on January 4, 1963, when it was 54,000,000 miles from the earth—a staggering distance! But plans for future steps in unmanned launch programs were even more staggering.

During much of this time, the majority at the spaceport gave undivided attention to their duties on Apollo. They knew that unmanned launches took place. In fact, they watched them go off regularly and had mostly unexpressed feelings of congratulations for Robert Gray and his most competent personnel in Unmanned Launch Operations.

Analyzing the Space Program

Less than two years had gone by since astronaut Gene Cernan and scientist-astronaut Dr. Jack Harrison Schmitt had walked on the moon. Five years had passed since Apollo 11 astronauts Neil Armstrong, Ed Aldrin, and Michael Collins had been the pioneers. They returned to complex 39 on July 16, 1974 to unveil a plaque commemorating the fifth anniversary of their pioneering journey. The inscription read in part: "Men began the first journey to the moon from this complex. The success of their explorations was made possible by the united efforts of government and industry and the support of the American people."[17]

Two years later, a group of analysts, led by Eli Ginzberg, made a study of public programs that concentrated on the NASA experience.[18] In the Apollo project, they pointed out, America spent approximately $40 billion in one decade with the specific mission of getting a man to the moon. It had done more than that. It had sent twelve men to the moon within twelve years. Even more important, it had brought the twelve back safely, as well as the three on Apollo 13 who circled the moon but, because of mechanical difficulty, had not landed.

In their first chapter, the Ginzberg Associates recalled that the legislation setting up NASA explicitly recognized that the agency's programs, like other federal programs, would influence the economic and social structure of the nation. The Rural Electrification Administration of President Franklin Delano Roosevelt's second term, for instance, did far more than bring electric lights to farmhouses throughout much of the contiguous forty-eight states. It introduced electric stoves, iceboxes, fans, mixers, electric milking machines, and mechanical lifts; brought stores to small towns to retail these appliances; and spurred industry to produce them. It cooperated with other agencies like the Tennessee Valley Authority and the Corps of Engineers.

The Ginzberg study asked five main questions: (1) Is NASA an old or new undertaking for the government? (2) What was the initiative for launching the program? (3) Did it have a specific limited goal or long-range and ongoing objectives? (4) Would the program carry on beyond the successful moon launches? (5) Were there any influential political networks?

To answer the questions in order, space exploration had not been a part of FDR's New Deal. President Truman had other priorities. Space exploration caught public attention only slowly. Shortly after the Soviets sent a satellite into orbit around the world on October 4, 1957, the National Aeronautics and Space Administration announced a manned space-flight program. When the Russians put Major Yuri Gagarin into orbit around the earth on April 12, 1961, the Soviet Union and the United States eyeballed each other in Cuba, in Berlin, and in space. President Kennedy challenged NASA to land men on the moon within the decade and bring them back safely. Clearly, this challenge called for a new and previously overarching achievement.

What spurred this unbelievable venture? The rivalry with the USSR on earth sent the USA into space. If that goal was reached, would the program carry on to new achievements? While NASA had a specific, clear-cut challenge to go to the moon within a specified time, it also had unmanned exploratory goals and had no intention of stopping space exploration if and when Apollo succeeded.

While the ultimate decision depended upon the leadership in Washington over the years, many factors worked to assure a positive response. A team of trained men was ready for similar ventures after Apollo. Man's curiosity to discover more about his universe was a significant factor and could well carry the program when hostility ceased between the great powers. The country enjoyed continued prosperity and could afford space exploration.

At the time of Apollo, a trio of Texans—Senate Majority Leader Lyndon Baines Johnson; Congressman Albert Thomas, chairman of the U.S. House Appropriations Subcommittee on Independent Offices; and Congressman Olin "Tiger" Teague, chairman of the Subcommittee on Manned Space Flight—formed that strong political network that the Ginzberg study considered necessary.[19] "When going got tough," men at the spaceport often remarked, "the 'Tiger' got tough." The Texas triumvirate could also count on an influential neighbor, Senator Robert Kerr of Oklahoma. Further, Johnson became president in crucial days for space exploration. Presidents Richard Nixon and Gerald Ford continued that support.

In short, NASA met the criteria that the Ginzberg team called for to ensure continuity.

The study indicated one negative aspect. It contrasted the centrality of the town of Huntsville, Alabama, in the operations at Marshall Space Center with the dispersal of population around Kennedy Space Center. But this was not a new discovery. When the men in charge of the Apollo program chose Cape Canaveral over Cumberland Island, they had recognized the one big weakness of the choice of Brevard: the lack of a central city as a focus of the space community. The people of Brevard had lived with this geographical situation and overcame any weaknesses arising from it.

6

WELCOMING A
"BENIGN BOOMERANG"

The Reusable Spacecraft

The first meaning of the word "boomerang" denotes an angular club that can be thrown so as to come back toward its starting point. Over the years, the word took on the connotation of an act or word that returns to haunt its originator. This second meaning has no place here. The first meaning dominates. The shuttle came back benignly, and opened a new path in space development.

After Apollo-Soyuz in 1975, a lull occurred in manned launches at KSC and in popular interest in space exploration. The men at the Cape regrouped and looked in new directions, especially to a reusable space-craft, called the Space Shuttle, planned by North American-Rockwell. In his book, *Beyond the Atmosphere: Early Years of Space Science*, Homer E. Newell saw its significance: "For a few years the diminishing urgency of the space program appeared to pose a threat to space science. But with the decision to proceed with the Space Shuttle, a renewed commitment to space science seemed assured."[1]

Newell believed that economy was the main factor in the shuttle's salability. "The country was not ready to support another costly space project unless it had some clearly foreseeable practical benefits. . . . It appeared that only by becoming much more efficient in the use of dollars could the space program continue in any shape comparable to that of the 1960s. That was perhaps the major issue in the years of discussion that preceded the final decision to build the Space Shuttle. The operational capabilities proposed for the new craft . . . captured the interest . . . and at least the tentative support of some leading scientists."[2]

Back in 1962, the Air Force had announced plans for a Dyna-Soar mini-shuttle to be launched by a conventional Titan 3C booster. A pool of six potential pilots was getting ready when headquarters canceled the project. Early in the next decade, President Nixon approved proposals for the Space Shuttle. In July 1972, NASA contracted with Rockwell International in Downey, California, to develop and test such a vehicle. Space visionaries made extensive plans that had to be streamlined or put off as time went on. Initially, the chosen design was far from fully usable. Eventually, engineers settled on three main elements: the Orbiter, the most complex flying machine ever built; a Delta-wing spacecraft about the length of a twin jet commercial airplane, but less sleek and built for a hundred flights; the external tank, built in the shape of a dirigible; and, attached to the tank's sides, two reusable solid rocket boosters, supplementing the engines. These looked like railroad tank cars standing on end. The three main shuttle engines had to provide 375,000 pounds of lift. They consumed a half-million gallons of liquid hydrogen in about eight and a half minutes. The solid rockets would put out a tail of flame 600 feet long and 200 feet wide. Planners thought they could get 100 flights from the Orbiter, fifty-five from the engines, and twenty from the major solid rocket booster components without large-scale refurbishment. The ultimate pattern had space for a shirt-sleeved crew of eight, with the possibility of another three. NASA planned to launch the shuttles from pads 39A and 39B. The shuttles took their names from great American sailing ships: *Columbia, Challenger, Discovery*, and *Atlantis*.

The crew of the old spaceport needed a new viewpoint. The spaceship they planned to send aloft, more sophisticated and complex than the Apollos, could return again and again to the space center for refurbishing and relaunching. By this time, many veteran space-workers of Mercury, Gemini, and Apollo days had retired. New names surfaced. One of them, Norm Albrecht of Yonkers, New York, had, as a member of the Air Force, attended the Jet Mechanics School in Amarillo, Texas. On completing his training, he served for some time at Westover Air Force Base in Massachusetts. In 1979, Rockwell International offered him the position of quality aircraft inspector on the Space Shuttle. The task thrilled him.

During the long months between Apollo-Soyuz and the arrival of the first Orbiter, engineers adapted the Apollo launch facilities to the needs of the shuttle. They modified the three mobile launchers in the VAB to stack and carry the new vehicle and changed the above-ground design of pads A and B on complex 39. The program needed an extended landing strip for Orbiter landings at KSC. In the meantime, the shuttle could land on the extra-long dry lakebeds at Edwards Air Force Base in California.

The men of NASA constructed a new two-bay building, called the Orbiter Processing Facility, near the VAB. A few years later, they built another building to refurbish the vehicle. The workers of NASA learned by doing and adjusted to the development as it unfolded.

The first Orbiter, *Columbia*, arrived at the spaceport in March 1979. The men of Kennedy Space Center, of Johnson Space Center, and of the contracting firms, took part in the 610 days of work at the Orbiter Processing Facility. Assembly and modifications were in order. Then the Orbiter spent thirty-five days in the VAB being mated with its booster rockets and fuel tank. It was out on the pad for five days before John Young and Robert Crippen went up on April 12, 1981. This first launch paled in contrast to Alan Shepard's first movement into space, John Glenn's circling of the world, or Neil Armstrong's and Ed Aldrin's landing on the moon, but it marked a new departure for NASA. Commander John Young had been in space more than any other American. A veteran of Geminis 3 and 10 and Apollos 10 and 16, he had walked on the moon on that last venture. Pilot Robert Crippen was going into space for the first time. No wonder the rate of his heartbeat soared just before lift-off.

The fact that the astronauts could bring the entire spacecraft back to earth for a number of future flights added a new thrill to the many previously enjoyed by the people of KSC. But this first launch revealed two diverse needs. The padmen needed new checkout and launch procedures, and the vehicle needed extensive rework. Technicians installed new sets of computers in the firing rooms of the launch control center. Crews soon learned to handle them.

Air Force Colonel Henry Engle and Navy Captain Richard Truly went up on the Orbiter *Columbia* on November 12, 1981. They too spent two days in orbit and landed in California. The first four-man crew of Commander Vance Brand, Pilot Robert Obermyer, Joe Allen, and Bill Lenoir rose in *Columbia* on November 11, 1982 and remained in space for five days. They too landed at Edwards.

In the meantime, Kennedy Space Center built its own runway northwest of the VAB, 15,000 feet long and 300 feet wide, four miles from launch pads 39A and B. In 1983, Crippen went up again, this time with Frederick Hauck, John M. Fabian, Dr. Norman Thargard, and the first woman in space, Dr. Sally Ride. As the *Columbia* floated several miles above the white clouds and blue oceans of earth, a free-flying German space platform, the SPAS-OI, took a photo of the open bay and of the sun-shield that protected two communications satellites before deployment.

The crew of *Challenger*, Vance Brand, Robert Gibson, Robert Stewart, Bruce McCandless, and Ronald McNair, landed at Kennedy on February 11, 1984, after almost eight days in space. The landing proved another joyous day for the people of the spaceport. Many KSC veterans long cherished the photograph of the shuttle touching down safely just four miles from pad 39A whence it had launched eight days before. The VAB dominates the right background of that popular photo.

Once the space shuttle mission was over, and the Orbiter *Columbia* back at Kennedy, the center again went into high gear. It launched one more space shuttle in 1981, three in 1982, four in 1983, five in 1984, and nine in 1985. Ten years before, no one would have thought such an achievement possible.

The twelfth shuttle, *Discovery*, placed another woman in space, Judy Resnik of Akron, Ohio. This flight landed at Edwards Air Force Base rather than at Kennedy Space Center. Later missions 13, 14, 15, and 16 came down at the KSC runway. By October 30, 1985, twenty-two successful missions had gone up. Concern grew for international cooperation in space. Representatives of various countries flew with the shuttle, among them Ernest Messerschmidt and Reinhard Furrer of West Germany, and Wubbo Ockles from Holland. The eight-man crew manned the first shuttle charter flight and the first in a series of space lab flights planned by West Germany.

The space shuttle lifted off like a conventional rocket, carried people, satellites, and experiments into earth's orbit, maneuvered in space, reentered at 18,000 miles per hour, and flew 1,500 miles through the atmosphere to a precision touchdown on a concrete runway. Using its robot arm and jet-powered flying armchair, American's men and women astronauts have carried out satellite retrievals and repairs and shown that they can do complex, heavy work in weightlessness. Compared with the crude throwaway capsules of the Apollo-Soyuz era, the reusable shuttle proved a miracle of flexibility and sophistication. It survived temperatures so high that they had reduced the Soyuz reentry module to little more than a blackened museum piece. The shuttle's main engines could fly again and again with perfect reliability. The landings were beautiful. The Orbiter came in like an underpowered glider at a speed of up to 226 miles per hour and at a glide angle of about 19°, more than six times steeper than that of a commercial jet liner when it lands.

The shuttle landing facility has a number of navigational and landing aids. Two recovery ships, the *Liberty Star* and the *Freedom Star*, recover the reusable solid rocket boosters after launch. The ships leave Cape Canaveral Air Force Station at hangar AF about twenty-four hours before launch and proceed to the predicted impact site. They cruise at a speed of ten to twelve knots and reach the area in about ten hours. In the time before lift-off, the ships conduct visual and electronic sweeps of the predicted impact area to ensure it is clear. Each ship recovers one solid rocket booster casing and three main parachutes. The ships return to hangar AF towing the solid rocket boosters, where they are cleaned and stripped before being shipped by rail to the prime contractor for refurbishing and reloading with propellant.

Weekends in Space

The space shuttle put a wide variety of individuals into orbit for long periods of time in a shirt-sleeve atmosphere. These included Gemini and Apollo veterans, and fifty new astronauts. Among them were Marc Garneau, the first Canadian in space, Senator Edwin Garn of Utah and Brevard's own Congressman Bill Nelson, Chairman of the House committee that oversees NASA. Congressman Nelson went up on January 12, 1986.

Congressman Nelson traveled with his committee and the Apollo-Soyuz astronauts to visit Moscow for the tenth anniversary of their linkup in space as guests of the Soviet Academy of Sciences. They talked about cooperation in space and how they looked forward to further cooperation with Russia. Coincidentally, in those same years that Congressman Nelson and the astronauts spoke of cooperation, the countries themselves began to look toward more friendly relationships at the end of the Cold War.

By 1986, NASA felt confident that it could launch a space shuttle every month. The vehicle had just started to fulfill its promise of frequent and economical access to orbit. NASA had launched a large variety of scientific and commercial spacecraft. The shuttle crew had repaired one expensive spacecraft in orbit. It recovered two large commercial satellites that had gone into improper orbit and returned them to the earth to be redeployed in space at a later date. The space shuttle was becoming a true national and international asset, carrying into space student experiments, small business payloads, and foreign payload specialists who tried their scientific experiments in orbit. The next major test for the space shuttle fleet—the assembly of a permanent manned space station—already had the approval of President Ronald Reagan and the Congress in January 1984.

There were four Orbiters in the fleet now—*Columbia, Challenger, Discovery,* and *Atlantis.* Then, on a relatively cold January 28, 1986, on the twenty-fifth shuttle mission, tragedy struck. The *Challenger* had carried forty-three astronauts in its nine previous missions, but, on its tenth mission, it exploded shortly after take-off from the Cape, bringing death to the crew. Up to this time, test pilots, scientists, and engineers traveled on the shuttle. *Challenger* carried Christa McAuliffe, "an ordinary citizen," as the words on her gravestone in New Hampshire declare, a teacher and the mother of young children. She, winner of a competition of over 11,000 applicants, rode the rocket to enlarge her worldview as a guide of the young. People all over the nation remember her name. They have forgotten the names of the other crew members, even that of Dr. Sally Ride, who had been the first American woman in space.

Veterans recalled the Apollo fire. "The day after the *Challenger* disaster," one remarked, "was like the day after Grissom, Chaffee, and White left us. What's next? We all wondered. Were we too cocksure, too complacent?"

Christa McAuliffe's words of encouragement before lift-off had recognized the ever-present danger of space exploration. She urged the nation to carry on should disaster occur. NASA resolved to do just that. But it did clamp down on security. Up to that time, for instance, I would go to Sheryll Pangborn of the security office badging desk, who would confer with an office at KSC, and Ms. Pangborn would clear my entrance. I had earlier taken the opportunity to view the new runway and other shuttle features.

On my next visit, I wished to interview three alumni of Parks College of Aeronautical Technology in Cahokia, Illinois, in gathering material for a life of the school's founder, the aviation pioneer Oliver Lafayette Parks. A security officer remained with me throughout the three interviews and then accompanied me to the front door. Casual visiting by former co-workers was no longer allowed.

The problems that caused the loss of *Challenger* and its crew required calm reflection and reassessment. NASA conducted the longest and most intensive investigation that it had ever run. The results showed an inadequate design of the fuel joints between the solid rocket booster segments. A NASA team redesigned these and retested them; checked other critical flight systems, reexamining and recertifying them at the same time; examined the launch facilities at the spaceport; and looked over the detailed procedures and software. Thirty-two months passed before everyone was satisfied that the vehicle and its crew were now as safe as it was humanly possible to make them.

On September 28, 1988, two years later, NASA sent *Discovery* into orbit. It was completely successful. The shuttles were shuttling again.

During the following year, 1989, Florida's governor and legislature created the Spaceport Florida Authority as a public corporation and subdivision of the state. It was responsible for the maintenance and improvement of Florida's position in space-related enterprises. It reaffirmed the state's many advantages: easy access to launch and support facilities, available financial assistance, a trade zone to alleviate import duties, an experienced pool of engineers, easy access to universities and research centers, a favorable tax environment, strong support from government on all levels, and general support of space industries.

By the end of 1995, space shuttles had flown forty-eight new missions, doubling the twenty-four safe flights that preceded the *Challenger* disaster. NASA had settled on a steady rate of seven or eight launches a year, and contracted for a new Orbiter, the *Endeavor*. It had its inaugural flight on May 7, 1992. The original promise of the space shuttle remained undiminished. The Kennedy-NASA industry team was doing its part to fulfill the dream of steady progress towards living and working on the new frontier of space.

The Spaceport News had carried a feature on James Jones, an engineer in KSC's Electrical and Mechanical Branch Facilities Division, who rode out a hurricane in a bunker at KSC thirty years before. In October 1994, the *Stargazer*, an in-house publication of Lockheed Space Operations Company, featured Jones again. His is one of the many individual contributions, otherwise unheralded, that space-workers have made to the outstanding achievements on the Cape. The writer, J. P. Kump, began:

> When Discovery finally came to life and lifted off the pad on Mission 6S7S64 early last month, it was after a nerve-wracking two-hour delay. The weather had everyone watching the skies for clouds and lightning. NASA engineer and Titusville resident Jim Jones was at his console in the launch control center wearing a wide smile on his face. Of course, Jones, a native of Zephyr Hills, wasn't the only one in the launch control center with a smile, but his came from a deeper appreciation of what it took to get that launch off in the face of those challenging weather conditions than most of the others. Jones, a systems electrical engineer for the past thirteen years in KSC's electrical and mechanical systems branch of the facilities division, helped design and install a critical environmental measuring system that allowed launch managers to make the right decision in that dynamic weather situation."[3]

Captain Jeff Lorens, a staff meteorologist with the Forty-Fifth Weather Squadron that supported NASA activities at KSC, believes the system Jones helped work on was crucial. "We would be at a significant disadvantage," Lorens stated, "without this system in trying to predict when and where lightning might happen. . . . It's a very valuable tool."[4]

Jones holds two patents in lightning measurement and tracking, and he has lived a long time with those staples of Florida summer weather, thunderstorms and lightning. A project engineer for the lightning field meter effort from 1966 to 1972, the thirty-year NASA veteran enjoyed the exciting challenge and the chance to work with some talented people. In the more recent past, Jones has also been instrumental in the installation, operation, and maintenance of the space shuttle visual landing aid system worldwide. In 1984, he traveled with a crew to Dakar in Africa to support a shuttle mission. The crew operated in some rather adverse conditions at the time. It was hard work with no vacation. By the turn of the century, Jones also answered NASA's call to work with the KSC energy working group to find new and better ways of conserving energy at the center.

A registered professional engineer in Florida, Jones holds a master's degree in systems management from the Florida Institute of Technology, and, like so many other parents in the area, he is involved in Scouts work and group activities. He and his wife Mary Ellen have three children: James, a twenty-seven-year-old Lieutenant and U.S. Naval Academy graduate and a veteran of Desert Storm; Mary Ellen, a twenty-five-year-old engineering graduate student at the University of Florida; and Kathleen, age nineteen, a nursing student also at the University of Florida. Two of his children are following in his footsteps, even though he said he made no effort to encourage them. A senior member of both the Institute of Electrical and Electronic Engineers and of the Instrument Society of America, Jones received NASA's certificate of recognition, and several group achievement awards. Like so many other KSC individuals, he takes part in community affairs. A competent toastmaster with the Toastmasters International, he has served as chairman and treasurer of Boy Scout Troop 481.[5] The space coast numbered many other similar achievers.

The ninety-seventh space shuttle launch and the fourteenth flight of *Endeavor* offered an opportunity to look more carefully at the totality of America's space mission. The purpose of this eleven-day jaunt into space was to produce unrivaled 3-D images of the earth's surfaces. The plan called for close to a trillion measurements of the earth's topography. It would contribute to better maps, improve water draining modeling, lead to the design of more realistic flight simulators, determine better locations for cell phone towers, and enhance navigation safety. Further, it might have some influence on flood control, soil conservation, reforestation, earthquake research, and glacier research.

It was a mission in joint partnership between NASA and the Department of Defense's National Imagery and Mapping Agency (IAMA), together with the German and Italian space agencies. The U.S. military is the primary customer of the data. It would use these 3-D pictures, called visualizations, to help in mission planning and rehearsal. During a minimum of 159 consecutive orbits, *Endeavor* mapped thirty-meter squares of the planet's surface at a time. The crew divided into three-person teams to conduct two work shifts of two hours each. Enough data would be available to fill 15,000 compact disks. Scientists would spend a year processing the massive volume of data.

A look at the background of the crew tells us much about the actual status of the space shuttle activities. The commander, Kevin R. Kriegel, was an experienced space flyer, who had been pilot on two previous missions and a commander on another. A U.S. Air Force pilot, the New York native accumulated more than 5,000 flight hours in thirty different aircraft. He joined NASA in 1990 as an aerospace engineer and instructor pilot, flying in the shuttle training aircraft and conducting the initial flight test of certain aircraft. He entered the astronaut program in 1992. He holds degrees in aeronautical engineering and public administration.

Navy pilot Dominic Gorie had gone up on an earlier shuttle. After high school in Miami, he earned numerous honors as a naval aviator and flew thirty-eight combat missions during Operation Desert Storm. He had earned a master's degree in aviation systems. Mission Specialist Janet L. Kavandi, a native of Missouri with a doctorate in analytical chemistry, had worked as an engineer in industry for a decade on space-related research. Her work on pressure-indicating paints resulted in two patents. Mission Specialist Janice Voss won a doctorate in electrical engineering from MIT. She joined the astronaut program in 1991 and worked on the space lab issues for the Astronaut Mission Development Branch. Mission Specialist Mamoru Mohri, Doctor of Chemistry and an expert on nuclear fission, had published more than 100 papers in the field of material and vacuum sciences. Selected for NASA's astronaut program in 1996, he joined the first group of exchange scientists in the U.S.-Japan Nuclear Fusion Program. Environmental physicist Dr. Gerhard P. J. Phiele, a German, had written extensively on physical and chemical oceanography. A member of the International Academy of Astronautics' Subcommittee on Lunar Development, Phiele joined the European space agency in 1998.

Endeavor truly boasted a remarkable crew!

The Shuttle Radar Topography Mission was the most ambitious earth-mapping mission ever. The ninety-ninth shuttle used two radar antennae, one located in the bay and the other on the end of a sixty-six foot deployable mast, to map earth's features. Its goal was to provide a three-dimensional topographic map of the world's surface up to the Arctic and Antarctic Circles.

Shuttle flights have resulted in the initial constructing and outfitting of the International Space Station; the orbiting of the Chandra X-ray Observatory; the repairing of the Hubble Space Telescope, and the mapping of the earth by means of high-resolution radar topography.

More lies ahead!

APPRAISING THE HOME FRONT

Brevard Bursts at the Seams

The census of 1980 showed Brevard with 272,959 people, of whom 68,034, or 25.7 percent, were under eighteen and 34,666, or 12 percent, were over sixty-five. The median age was thirty-four years and one month. One hundred and three thousand or 43 percent of the 273,000 citizens of Brevard were under twenty-five. Such statistics would have raised eyebrows anywhere, but especially in Florida where more and more elderly were arriving every year and the population approached ten million, a 43 percent increase over the census of 1970.

Over half the county's population, roughly 57 percent, resided in six communities. Melbourne led the way with 46,536 residents. Titusville followed with 31,910. Merritt Island ranked third with 30,708. Palm Bay, destined to soar in later censuses, already had 18,560. Cocoa numbered 16,096 and Rockledge 11,877.

Five thousand Brevard businesses paid their 80,000 workers $1 billion a year. Nine firms hired 1,000 or more workers, a dramatic change from the unique position of Harris Electronics as the only major industry at the start of the spaceport. Two hundred and forty-three manufacturers employed slightly fewer than 25,000 workers with a payroll of $110 million. That represented a growth of 139 new firms during the 1970s. The five guided missile factories hired almost 5,000 workers, with slightly over $100 million in annual payroll. But more expansion lay ahead.

The decade of the 1980s saw the number of businesses almost double from slightly over 5,000 to 9,744. While 81,503 Brevardians had worked in 1980, 136,686 had jobs in the 1990s, with an annual payroll of almost $3 billion. Contractors changed the face of the county. From a combined total of 627 in 1980, their numbers rose to 1,456: 667 in heavy industry construction and 789 in home building. While 243 manufacturing plants flourished in 1980, by the mid-1990s almost twice that many had work for 25,180 employees and met an annual payroll of over $700 million. Service industries had more than doubled in number from 1,467 in 1980 to 3,301 in 1990. They hired 53,023 employees and their payroll reached $1,303,094,000. Health services continued to swell, more than doubling in extent of institutions and personnel.

By that time, the abundance of fast foods on U.S. 1 made the casual visitors wonder if any Brevardians ate at home. Residents had their choice of 735 places to eat, 276 food stores, 56 pharmacies, 129 apparel shops, 220 furniture stores, 186 service stations, and 111 building material and garden supply stores. What a far cry from the early days of the spaceport when Jim Deese, Wally Dal Santo, and Ernie Reyes had to travel to Orlando for basic family needs!

The median income in Brevard County in the 1960 census, $6,123, had been the highest for any county in the state. By 1990 it had risen to $30,534, but fell to sixth in the state, not far behind Seminole County.

To put the Brevard County economic situation into focus, a comparative look at one similar and one dissimilar county will help. While Volusia County, immediately to the north along the Atlantic, had only one city of size, Daytona Beach, with 61,000 inhabitants, the county had a population of 370,712, almost as large as Brevard's 399,978. Volusia surpassed Brevard in the number of retail establishments and the size of its payroll, but had welcomed only half as many manufacturing concerns: 11,204 to Brevard's 22,702. The payrolls, too, stood apart. Volusia's manufacturers paid slightly under $500 million in wages while Brevard handed out $863,783,000.

Both these counties stood in sharp contrast to St. Johns farther up the Atlantic coast, around St. Augustine. In this heavily tourist and retirement area 32,000 people lived. Retail sales were high in St. Johns, but only 2,510 firms manufactured goods. Brevard and Volusia completely overshadowed St. Johns in many ways. But the people of the latter county developed far more senior citizens' programs than the other two.

By this time, sons and daughters of space-workers looked to employment in the program. Wallace Dal Santo's daughter hoped to enter astronaut training. Both sons of Art Gruenenfelder followed their father to the Cape. David Gruenenfelder started with Lockheed and later worked on air conditioning of vehicles and buildings. His wife, Kim Gardener, also worked at the Cape in radar. Art Jr. "Bud" Gruenenfelder was, like his brother, in air conditioning. Bud's wife Linda worked at the VAB as a computer analyst. David Gruenenfelder reports, "I feel we're part of history." The new generation carried on in the spirit of its predecessors.

When Jim Voor and his wife Betty came to KSC in 1965, they had three boys and two girls. The family grew to include five more children during its early years in Florida. In the late 1980s, Voor mentioned to his boss that he had four sons working at KSC.

"That's nepotism," the boss said.

"They're all working for different companies or agencies," Voor responded. The next day he found that McDonnell-Douglas had hired a fifth son. Later on, his daughter Cindy worked at KSC for the National Wildlife Refuge. Joe, a star football player at Astronaut High School—on the same team as Cris Collinsworth, the All-American star and later a professional player—worked at the Johnson Space Center. Of the other Voor children, Andy was a technician in non-destruct testing and evaluation at KSC. David worked in Orlando in Naval Civil Service. Tom at Motorola designed cellular telephones, and Robert worked in computer telephony in Tampa. One daughter's husband, Steve, was in the computer and electronics field. Donna's husband James was a supervisor for U.S. Alliance at KSC. Jim, a thirty-five-year veteran, still commutes to KSC five days a week.

Space seemed to be a Voor family enterprise.

Women were going into space and into more significant places at KSC. Dr. Sandra Magnus, for instance, joined NASA's astronaut corps. This native of Belleville, Illinois, had taken her bachelor's degree in physics and a master's degree in electrical engineering at the University of Missouri at Rolla before working for McDonnell-Douglas. She received a doctorate in science and engineering from Georgia Tech. At the time of this writing she was responsible for communications between the crew of the international space station and the controllers on earth. She had not yet been in space, but looked forward to spending time on the space station.[1] Her experience gives another example of the growing welcome that women engineers were receiving—in contrast to the early days at KSC.

Another change in the employment pattern came during those late years of the century. None of the Mercury, Gemini, or Apollo astronauts had taken positions at KSC. Among these great achievers, John Glenn and Dr. Jack Harrison Schmitt served terms in the U.S. Senate. William Anders became Ambassador to Norway. Charles Conrad was an ambassador of good will for McDonnell-Douglas Corporation. Neil Armstrong held a professor's chair at the University of Cincinnati. Some shuttle veterans changed that pattern. Donald R. McMonagle, for instance, a veteran of three shuttle missions between 1991 and 1994, answered NASA's call to set up a new Extra-vehicle Activity Project Office. In this position, he managed the resources and planned and oversaw the space walks in support of the space shuttle and the International Space Station.

In the following year, McMonagle managed launch integration for the space shuttle program. In that position, he prepared for shuttle launch, oversaw the launch, and, if the vehicle landed at a remote location, ensured safe return. He also served as chairman of the Space Shuttle Mission Management Team for launches. "The team that processes launches and flies the space shuttle is composed of the most dedicated people with whom I have ever worked," McMonagle said, "and I have been deeply honored to count myself among them. The space shuttle today is safer than it has ever been, and I am confident that the space shuttle team is more firmly committed to achieving excellence in their job now than ever before."[2]

Silver-Haired Citizens

An entire generation of space-workers retired in the area. No matter where they came from—Long Island, St. Louis, Omaha, Seattle, or wherever—they chose to stay. Even those who had lived and worked a time in California remained. The area offered so many advantages: more sun than Seattle, less snow than New England, beaches, boating, and memories of great achievements.

Further, newcomers in great numbers were moving in to enjoy the advantages of the area. "Over-65s" reached roughly 16.1 percent by the time of the 1990 census.[3] Problems arose, too. Gone were the neighborhood shops, the mom-and-pop stores they knew in Brooklyn or St. Louis. Groceries, bakeries, hardware stores, travel agencies, and pharmacies were clustered at shopping centers. Fast food and other restaurants lined the highways. Every family depended on a car. When everyone in the household passed the age of safe driving, a new way was needed to meet basic needs.

Many oldsters needed to rely on their neighbors for access to malls, to churches, to doctors' offices, and to programs of various kinds. Some of these needs, such as transportation, could be provided by neighbors. Church groups could help with others. Some required community awareness, and newspapers could make people aware of neighborly needs through editorials and the letters-to-the-editor column. Numerous seniors lived in mobile homes on rented lots. Sometimes owners gave them only thirty days to move. To resolve such problems, governmental action on the local, state, or national level was needed. Help came on the state level.

A group of elderly Floridians came to look with approval upon an organization called the "Silver-Haired Legislature" that began in Missouri. While Missourians have the reputation for preferring the proven idea, thanks to their slogan, "You've got to show me," Show-Me Staters have actually pioneered in many areas, most recently in choosing the design of Aero Saarinen, the Finnish architect, for the Jefferson Memorial Gateway Arch, the unchallenged monument of the iron-and-steel age. Floridians were happy to follow Missouri's lead.

The Florida unit departed, however, in one major way from the Missouri pattern. While the latter met during the legislative session in Jefferson City, Sunshine Staters gathered in Tallahassee at other times. As a result of this arrangement, leaders could meet in the official chambers at the heart of the state government. More and more people became aware of the needs of oldsters.

Bob Wattels, a lawyer from Orlando, served as first executive director. Senator Claude Pepper gave his strong but quiet support. Brevard was well represented. Ralph Hall, schoolteacher and later dean, became executive director between 1988 and 1992. Former Titusville Mayor, Vern Jansen, whose experiences have already enriched this book, also represented his fellow Brevardians.

Ralph Hall was born in Manderson in North Central Wyoming, a place known only to God, Shoshone Indians, countless antelopes, and an occasional Easterner lost on his way to Yellowstone Park. Ralph remembers the cold of the countryside, with an occasional twenty-four feet of snow. When he left Wyoming, with its relentless winds, deep snows, and intense cold, to attend Ohio State University, and then MIT, he kept looking toward warmer climates. He served in the diplomatic service in Malaysia in the early 1960s, in Vietnam from 1962 to 1965, and in Laos from 1965 to 1967. After these many years in Southeast Asia, it took him a long time to adjust to the quiet routines of life in America. He adjusted immediately, however, to the warmth of sunny Florida. He became dean of a public school and taught science at the junior high level at Rockledge and at Astronaut High in Titusville.[4]

Vern Jansen never walked through deep snows at Quincy, Illinois, but he did not like the area as did his wife, Bernice. Immediately after World War II, he worked at Fort Monroe in Virginia, and then at Anniston Ordnance Depot in Alabama. He came to the Cape in 1951 and later worked for NASA.

As mayor of Titusville in 1969 and 1970, he witnessed the layoffs that came shortly after the successful flight of Apollo 11. He kept a steady hand on the controls during the time of adjustment. After his term, while continuing to work at KSC, he promoted Scout programs, especially projects Scouts undertook to win their Eagle badges.[5]

Hall and Jansen and their associates in the Silver-Haired Legislature knew that many issues had to be faced. It was good to have a group of citizens meeting to call attention to these issues and suggest remedies, such as arranging for seniors to renew their driver's licenses by mail.

A major achievement stemmed from calling attention to an omission in health care training at the medical schools of the state. Even though Florida boasted four medical schools, one each at the University of Florida, the University of Miami, the University of West Florida, and the University of South Florida, none had geriatrics programs at the time. In response to the efforts of the Silver-Haired Legislature, two universities moved in this area.

The *Golden Year Magazine*, edited by Sid Caplan, told the story of these programs. Both Vern Jansen and Ralph Hall felt that the many good things they were able to do for the senior citizens made the Silver-Haired Legislature a really important program in the development of Florida.

Living in Brevard

Brevard County looked to the third millennium with confidence. Central to the dynamism of the area, KSC moved confidently. The center controlled about 140,000 of Brevard's acres. It owned 84,000 of those and, by agreement with the Department of the Interior, administered the rest for the State of Florida as a wildlife refuge. It included a long stretch of seashore. The United States sent up all its manned space flights from here and such unmanned probes and explorers as the Viking to Mars.

KSC launched the Space Shuttle and built a long runway for its return. The area held responsibility for the assembly, the checkout, and launching of the vehicles and their payloads, the loading operations, and the turnaround of the Space Shuttle Orbiters between missions. It prepared for continuous launches of the components of the space station in the new millennium, and will remain the only site for U.S. manned space flights for the foreseeable future.

During the fiscal year 1998, KSC's space-related employment and contracts boosted Florida's economy by $966 million. Of this, NASA spent $799 million in Brevard County. Of this, $762 million went directly to contractors operating at KSC, and $37 million went to office site businesses located in Brevard County. Among the major contractors, United Space Alliance handled the shuttle, Space Gateway Support Florida Incorporated supervised base operations, the Boeing Company oversaw payload ground operations, and Dynacs directed the engineering development.

The Forty-Fifth Space Wing hosted Patrick Air Force Base, the Cape Canaveral Air Station, the Florida instrumentation stations, and island stations of the eastern range. That range stretched from Cape Canaveral, 4,700 miles southeast to Ascension Island in the South Atlantic. The Forty-Fifth Space Wing operated the eastern-range space launch facilities and the instrumentation that checks conditions in space and rates the performance of ballistic missiles.

Patrick Air Force Base, lying three miles south of Cocoa Beach and covering 2,100 acres between the Banana River and the Atlantic Ocean, housed the headquarters of the Forty-Fifth Space Wing. The Cape Canaveral Air Station, twenty-one miles north of Patrick Air Force Base, had served as a launch site for the nation's exploration of space for fifty years and supported over 3,200 launches. Within Cape Canaveral Air Station's 16,000 acres, the Air Force developed complete assembly and launch facilities for ballistic missiles and space hardware. Those in command could track and monitor flights and destroy any errant missile that threatened public safety. A 10,000-foot landing strip permitted the airlift of missiles directly from the place of manufacture to the Cape Canaveral Air Station.

The eastern range, the nation's primary launch site for commercial space programs, has been on the forefront of policy development. In this, it cooperated with all levels of command within the Air Force, the Department of Defense, and the Office of Space Transportation for the Federal Aviation Administration. Its list of users included NASA as well as the Atlas, the Delta, the Titan, the U.S. Navy Trident, the United Kingdom Trident, and the Athena. The area anticipated an average of thirty-four launches, almost three a month, through the succeeding five years.

The Forty-Fifth Space Wing numbered 8,679 individuals, service and civilian, on its $463.9 million payroll. It generated $1.44 billion in business in Brevard County. Of that, $762 million went to support contracts, $28 million to construction, and $12.3 million to other procurement. Financial experts estimated the value of jobs created at $172.5 million. The Space Wing also had accommodations for maintenance units, such as the Air Force Technical Application Center and the 920th Rescue Group that brought the number of employed personnel to over 10,450. The area also welcomed a large population of retired military personnel.

While Harris Corporation in Melbourne was the only employer in Brevard with over twenty employees at the start of the space age, and, by the year 2000, numbered 6,000, twenty-one firms employed more than 500 employees. The Boeing Company, that worked with the Space Shuttle Orbiter vehicle, employed 2,900; Brevard Community College 1,812; the Brevard County Board of Commissioners 2,592; and the Brevard School Board 7,915.

Six hundred journalists and others worked for Cape publications. One thousand, one hundred Raytheon technicians operated and maintained instruments on the eastern range. Dictaphone enrolled 599 in various capacities. Eight hundred and fifty Excell Agent Services personnel in Rockledge provided directory assistance. Florida Tech had a payroll of 700. Four thousand three hundred worked for Health First, Inc. Intersil Corporation kept 1,600 busy in Melbourne. In the town of Cape Canaveral, Johnson Controls World Services Corporation engaged 500. Lockheed-Martin, a combination of spaceport veterans, still employed 1,000 skilled personnel. Over 1,600 worked for NASA itself. Northrop-Grumman Surveillance & Battle Management Systems employed 2,000; the Parrish Medical Center, a non-profit hospital in Titusville, slightly over 1,000; Rockwell Collins 1,100; Sea Ray Boats Incorporated, a boat manufacturer, 700; Space Gateway Support, another NASA contractor, 2,663; Sverdrup Technology, another NASA contractor, 500; the United Space Alliance, NASA's space flight operations contractor, 6,000; and the Wuesthoff Health Systems, 2,018. Six hundred skilled workers manufactured circuit boards at Tyco Printed Circuit Group in Melbourne.

Besides major airports at Orlando and at Melbourne, small airports served the area: Arthur Dunn Airport in Titusville, Merritt Island Airport, Space Coast Regional Airport in Titusville, Titus-Cocoaville Airport Authority, and Valkaria Airport in Palm Bay. A number of cruise lines went out from Cape Canaveral: the Cape Canaveral Cruise Line, the Carnival Cruise Line, the Disney Cruise Line, Premier Cruise Line, the Royal Caribbean International, Sun Cruise Casino, and Sterling Casino Line.

In place of the limited hospital facilities in the years after World War II, the county boasted Cape Canaveral Hospital near Cocoa Beach, Holmes Regional Medical Center in Melbourne, the Omni Urgent Care Center, the Palm Bay Community Hospital, the Parrish Medical Center in Titusville, and the Wuesthoff Hospital in Rockledge. Five solid Chambers of Commerce offices served Titusville, Cocoa Beach, Melbourne, Palm Bay, and Sebastian River. A north-south route of the Greyhound Bus Lines connected the area from Titusville to Melbourne.

Of a labor force of 204,000 in the county, only 8,000 were unemployed. The average rate of unemployment for Florida was 4.3, slightly less than the national mean of 4.5. The average salary in 1998 ranged from $43,000 in mining and $39,000 in manufacturing to $16,000 in agriculture and $14,000 in retail trade. That indicated a median of $28,000.

Opportunities for outdoor recreation were many. Brevardians had access to 145 parks, with forty of them in District 1 of the northern area, among them Fox Lake Park and Singleton Tennis Complex; twenty-seven in District 2 North; nineteen in District 3 of the southern area, including a complex named for Flutie, the Boston College football hero who beat Miami; seventeen in the central mainland area, especially Rockledge Park; and seven in District 5, featuring Paradise Beach Park.

Eighteen tennis courts were scattered throughout the county, three in Titusville, four in Melbourne, two in Rockledge, two in Indian Harbor Beach, and one each in other areas. There were twenty-four golf courses and seventy-two nature trails, seven in the north, nine in central, and six in the south. The Brevard Zoo had 400 animals. In conjunction with it, one could visit the Wetlands Outpost, a 22-acre preserve, either by kayak or on a guided tour.

Brevardians had beaches for swimming and relaxing, docks for boating, and swimming pools, public and private. Major college football was within driving range with the University of Miami, Florida State, and the University of Florida regularly challenging the best in the nation. What made the games even more interesting over the years was that spaceport workers brought their loyalties with them. Chicagoan Jack Tobin still rooted for Notre Dame after years in Florida. Jimmy Bronwell's devotion to the Crimson Tide of Alabama surged long after he moved from Huntsville to KSC headquarters. The Tampa Bay Buccaneers shared the professional spotlight with the Miami Dolphins for a few years. The Jacksonville Jaguars joined them a few years later. All three places were no farther away than the distances Green Bay fans had been driving to see their revered Packers. Younger Brevardians tended to give their loyalties to Florida teams. Not so the older generation. Charley Wingertsahn remained loyal to his hometown Pittsburgh Steelers after working for thirty-five years with an electrical firm in Titusville. Boeing veterans still cherished hopes that someday the Seattle Seahawks might come through and win the Super Bowl.

Brevard County had its own hero on the gridiron, Cris Collinsworth of Astronaut High in Titusville, the University of Florida, and the Cincinnati Bengals. A fluent and clever speaker and an excellent play analyst, Collinsworth remained in the football spotlight as analyst on NFL football reviews.

Schooling in the County

America pioneered the idea that all could learn, and our forefathers set out to make education available to all its citizens. Brevard accepted that challenge. It took for granted that everyone can learn, not necessarily with the same facility or success, but with results that could benefit the individual and society. It recognized individual as well as cultural differences and looked on them as enriching American life. The ninth largest school district in Florida, Brevard's public school system, ranked forty-eighth in size among the more than 16,000 school districts in the nation during the 1990s. The district owned and maintained ten million square feet of facilities on more than 2,000 acres of land. An estimate of the value of the facilities and property approached $800 million.

Of these schools, fifty-four elementary schools taught 37,124 students; fourteen middle schools taught 10,933; and twelve senior highs served 20,362. Eight charter schools served 1,216 students, and 16 special schools taught 794. In all, 70,429 students attended 104 schools. Of those attending, close to 80 percent were Caucasian, 14 percent Black, almost 5 percent Hispanic, less than 2 percent Asian, and 2 percent from other ethnic groups.

Three Brevard schools—Gardendale, Riverview, and *Challenger* 7—were on a year-round schedule. There were six elementary magnet schools: Ronald McNair and Endeavour in Cocoa; Gardendale in Merritt Island; Gulfview in Rockledge; University Park in Melbourne; and Riverview in Titusville. Three elementary schools of choice were scattered in various places, with Freedom 7 School of International Studies in Cocoa Beach; Robert Louis Stevenson School of the Arts in Merritt Island; and West Melbourne School of Science in Melbourne. Also in Melbourne, Westshore Senior High School furnished a secondary school of choice. Cocoa Beach Senior High School offered the international baccalaureate program for academically talented students.

Eight public charter schools also offered programs for various grade levels.

The state law required that students living beyond two miles of their school be provided transportation to school. By the year 2000, 365 buses transported 28,000 students daily. The Brevard School District Transportation System, one of the largest in Florida, traveled about six million miles a year, or 36,000 on an average school day. The transportation operating cost went slightly over $13 million. The state funded about 70 percent of that figure. Brevard County spent slightly under $5,000 per pupil each year.

Teachers' salaries ranged from $27,000 for those with only a bachelor's degree to $32,000 for those with doctorates. The average teachers' salary with benefits was $37,000. The fringe benefits totaled almost $10,000. The total annual school budget rose to $550 million. Sixty-six percent of that came from the state, 33.36 percent from the local taxes, with little from the federal government. Most of the expenditures were for salaries and fringe benefits.

In looking at this school system, one must recognize, as did Newton Gregg cited in an earlier section of this book, that Brevard County boasted an exceptional number of college graduates among concerned parents. The whole atmosphere went far beyond that of the average county in the United States. Few counties in the nation could match it.

The Brevard public schools set out to offer a quality education to every student in a safe, secure, orderly, and healthy environment. In the light of the Columbine outburst in Colorado, Brevard's insistence on safe educational surroundings reflected a national concern. But Brevard worked to that goal not with a call to police to patrol the school halls, but with a challenge to students, parents, educators, and the community at large to "shape up." Brevard taught its young in a dignified and professional manner and expected a dignified response. When it called on parents to involve themselves in the education of their children, it won an enthusiastic response.

The record of graduates of Brevard high schools who won nominations to the U.S. Air Force Academy in Colorado Springs attested to the quality of Brevard secondary education, public and private. Dr. Kathleen O'Donnell, chief of the Cadet Operational Analysis Office at the academy, pointed out the large attendance of cadets from Brevard. Fifty-seven appointees have come from the county since 1970: three each from Palm Bay and Cocoa Beach, five from Cocoa, six from Titusville, eight from Rockledge, and thirty-two from Melbourne.

The first Brevardian to graduate from the Air Force Academy was Henry P. Mitchell of Cocoa Beach in 1970. During the school year 2000–2001, Jonathan D. Haun of Cocoa, Jason S. Rogers of Titusville, Natalie R. Vincent of Rockledge, and four from Melbourne—Shaun A. Johnson, Robin E. Lease, Jay Rubica, and Michael G. Opresko—pursued their courses at the foot of the Rockies.

Up to the present, three women cadets, all from Melbourne, have graduated from the Air Force Academy: Ann Marie Matonak of the class of 1981, Wanda Kay McCoy in 1983, and Teresa Draughn in 1986. The largest single class of Brevardians, the graduates of 1979, numbered seven. Between 1972 and 1996 the county had at least one representative in each graduating class. In all years since 1970, the U.S. Air Force Academy included at least one cadet from the space coast.[6]

Commuter Colleges

Brevardians attended colleges and universities throughout the country as well as in other parts of their state, especially the University of Florida at Gainesville. Nonetheless, the county itself and neighboring areas offered countless opportunities for higher education. An earlier chapter told of the origin and development of Brevard Community College and Florida Institute of Technology.

Brevard Community College had educated 30,527 students as it approached the end of the century. With campuses at Cocoa, Melbourne, Titusville, and Palm Bay and centers at KSC and Patrick Air Force Base, it covered the county. The Melbourne campus of BCC featured the Maxwell C. King Center for the Performing Arts and a 2,000-seat theater. The Henegar Center for Arts in Melbourne had a seating capacity of 500.

The Florida Institute of Technology embraced five separate schools by 2001: the College of Science and Liberal Arts, the College of Engineering, the School of Business, the School of Psychology, and the

School of Aeronautics. These offer a full scope of programs leading to degrees at the bachelor, master, and doctoral levels. The School of Extended Graduate Studies offers master's degree programs at ten graduate centers in five states. The Flight Operations Department of the School of Aeronautics is located at the Melbourne International Airport, five minutes from the campus.

Also of great importance for the spaceport collegian, though not in Brevard County, the University of Central Florida in the eastern section of Orlando owed its foundation to the efforts of the authorities at KSC and Patrick Air Force Base. A member of the State University System of Florida and accredited by the Southern Association of Colleges and Schools, it grants the associate, bachelor's, master's, and doctoral degrees. For a relatively young school, it has an amazing collection of books—1,199,253 hardbound volumes. Classes began in 1968, with 1,492 freshman and junior students. By the year 2001, 17,203 undergraduates and 5,055 graduate students attended full-time and 7,948 part-time. It opened a satellite campus in Cocoa.

One college and four universities, situated elsewhere, sponsored centers in Brevard: Orlando College, and Barry, Embry-Riddle, Webster, and Florida Metropolitan universities. Florida Metropolitan opened one of its eight campuses in Melbourne. Orlando College, a private business college approved by the Accrediting Council of the Association of Independent Colleges and Schools, operated on a quarter system and offered one summer session. It had an enrollment of 1,700 students. While the basic campus was in Orlando, it opened a branch campus in Melbourne.

Barry University of Miami Shores, an independent coeducational Catholic institution of higher education in the liberal arts and professional studies within the centuries-old Dominican tradition, was founded in 1940 and became a university in 1981. Accredited by the Southern Association of Colleges and Schools, Barry granted degrees at the bachelor, master, and doctorate levels. It had nine Florida off-campus centers, including one in the area of the spaceport.

Webster University in suburban St. Louis grew out of a splendid women's college of the same name, organized and developed by the Sisters of Loretto, members of an institute of American nuns founded in Kentucky in 1815. In the period of the 1960s, administrators moved from a religious-oriented college to an independent liberal arts university with no religious affiliation. It developed programs at a number of campuses, many of them connected with Air Force installations. It enrolled 13,168 students worldwide, while the undergraduate college in the St. Louis metropolitan area registered 1,885 full-time and 995 part-time undergraduates, plus 2,372 graduate students. In 1993 Webster opened centers at Merritt Island and Palm Bay.

Embry-Riddle Aeronautical University of Daytona Beach, one of the oldest aeronautical schools in the country, operated a center at Patrick Air Force Base that enrolled 1,200. It offered bachelor's and master's degrees in aeronautical science, management, and associated areas.

Within easy driving range, roughly the distance many Los Angeles collegians drove daily to UCLA or Loyola-Marymount, four institutions welcomed Brevardian commuters: Rollins in Winter Park, Stetson in DeLand, Daytona Beach Community College, and Bethune-Cookman, also in Daytona Beach.

Rollins College in Winter Park, Florida, a private liberal arts college founded in 1885, was the first to offer college-level work in Florida. Accredited by the Southern Association of Colleges and Schools, it awarded the bachelor and master's degrees and offered continuing education programs. It was convenient to the spaceport community.

Stetson University in DeLand won its name from a benefactor who made his fortune developing that great hat that bears his name. It is a private, comprehensive university, accredited by the Southern Association of Colleges and Schools, with a College of Arts and Sciences and Schools of Music and Business Administration.

Bethune-Cookman College in Daytona Beach is a private, coeducational college affiliated with the United Methodist Church and accredited by the Southern Association of Colleges and Schools. Bethune was founded in 1904 and later combined with Cookman College. At that time, it provided much-needed rudimentary training for African American boys and girls and later welcomed students of other races. By the end of the century, it had an enrollment of 2,292 full-time and 189 part-time students, a faculty of 133 full-time and eighty-two part-time instructors with a good student-faculty ratio of seventeen to one.

The Daytona Beach Community College was not too far away, and its south campus in New Smyrna Beach was even closer to Brevard County. Authorized in 1957 by the Florida Legislature, it became the state's first comprehensive community college, accredited by the Southern Association of Colleges and Schools, and offering multiple academic opportunities, including courses in seven foreign languages. By the end of the century, the yearly enrollment was 36,500 students in all programs, with 17,490 students in college credit courses. Undergraduate degree enrollment totaled 4,608 full-time and 6,553 part-time students. The student-faculty ratio was twenty-two to one. All Brevardians who wished could get a college education close to home.

The tremendous growth of opportunities on the college and graduate levels and the vigorous response of young people to these possibilities points to the background of most of the Brevardian families. On the one hand, Brevard had no decaying inner-city and no large body of unskilled workers, as did so many highly populated areas of the country. On the other hand, it numbered few families where neither parent was college-educated. And these few wanted a college education for their children as much as the graduates of Georgia Tech or Georgetown. Further, the wide regional interest in attending the services academies, Army, Navy, Air Force, or Coast Guard, met few matches anywhere in the country.

Attendance at college was a taken-for-granted way for young Brevardians.

LOOKING BEYOND
THE FAR HORIZONS

Robots Plot the Planets

Without the fanfare or cost of Apollo or the shuttle, relatively inexpensive unmanned launches had continued to move through the vast universe and to send back valuable information.

Engineers designed Pioneer 10 to fly by Jupiter and Pioneer 11 to track down Saturn. The former went up on March 3, 1972, the latter on April 6 of the following year. A year and a half later Pioneer 10 approached Jupiter and studied that body. Pioneer 11 imitated its sister vehicle to make a similar survey and rendezvoused with Saturn and its moon Titon. Later on, it became the first spacecraft to visit Uranus.

On November 3, 1973, Mariner 10 went up from Cape Canaveral aimed at the planet Mercury. With a wingspan of twenty-five feet and equipped with solar panels, it bypassed Venus at a distance of 3,600 miles and photographed the cloud tops. Going by Mercury on March 29, 1974, less than two months after bypassing Venus, it sent back pictures from a distance of over 3 million miles. Mariner 10 eventually entered a permanent path around the sun that brought it back to the neighborhood of Mercury on September 12. It sent excellent photographs and television pictures. The mission ended on March 16, 1975. By March 24, all contact was lost and to all scientific purposes, Mariner was dead. But it still circled the sun and came near Venus every six months. Reports on the hard-to-believe distances and the incredible speed of the vehicles moving through space continually awed the nonprofessional observer.

Back in 1961, President Kennedy's first call had been for a moon landing in the decade. His last call, given at the spaceport less than a month before he died, called for international cooperation in space. By the mid-1970s, the once–Cold War enemies worked simultaneously to advance man's knowledge of his universe.

During October 1975, Russian spacecraft Venera 9 and Venera 10 came down on the surface of Venus and sent back pictures. The amount of light available was more than expected, about equivalent to noon light in Moscow on a cloudy January day. The U.S. countered with Venus Mission 1 that was to send data for a decade.

Venus Mission 2, a complicated contraption, crash-landed on Venus. One of the probes survived the crash and sent data for sixty-seven minutes. The Russians kept at it with Veneras 13, 14, 15, and 16 in the early 1980s.[1] By that time, mankind had gained good maps of Venus's surface. The Infrared Astronomical Satellite (IRAS) in 1983 made the first detailed examination of the universe. It discovered new stars being born and possible evolution of new planetary systems.[2]

Many countries involved themselves in two Vega dual missions to Venus and Halley's comet in December 1984: the U.S., the USSR, Austria, France, Bulgaria, Czechoslovakia, Hungary, East Germany, West Germany, and Poland. Two spacecraft went up in that month and flew by Venus four days apart the following June, 1985. Each carried a lander that did not work as efficiently as hoped, but did travel more than 6,200 miles over the surface of the planet and sent back information on the composition of the soil. On the second half of their mission, the Vegas sent back views of Halley's comet.

NASA's *Magellan*, lauched from the shuttle *Atlantis* on May 4, 1989, was equipped with instruments that were able to map Venus through its cloud layer. Space historian Tom Crouch described this breakthrough: "Venus, the great mystery of the solar system since the invention of the telescope, was transformed into one of the best known and most thoroughly mapped objects in the solar system."[3]

While each shuttle seemed to match its older brothers, each unmanned launch became more sophisticated. The *Galileo* spacecraft, launched by the shuttle *Atlantis* on October 18, 1989, took a long and energy-efficient path to the moon. It picked up energy on the way by means of a Venus-Earth gravity assist trajectory. It seemed to delight in bouncing around the universe, sweeping close to Venus, swinging around the sun, flying twice near the earth, and then reaching for Jupiter in its seventh year.

Twenty years had gone by since the Viking lander went down on Mars and showed the great possibilities of planetary exploration by robot. Man had found no life anywhere but on his own home, the planet Earth. On July 4, 1997, a Mars *Pathfinder* celebrated Independence Day by coasting down through the thin atmosphere and bounced onto the surface of the Red Planet. Inflatable shock-absorbing balls cushioned the landing. It sent out weather reports and photographs of the area. A typical photo showed terrain that was not likely to spur a Kansas farmer to plant spring wheat. The Sahara seemed more able to sustain life.

As a special feature, this spacecraft carried a twenty-five-pound, six-wheeled robot explorer. It bore the name *Sojourner*, in honor of Sojourner Truth, the heroic African American advocate of justice and freedom in the days of slavery. The roving *Sojourner* and the stationary *Pathfinder* spent the rest of the summer and early fall working on three eight-hour shifts.

"During an active life of four months," space historian Tom D. Crouch writes in *Aiming for the Stars*, "the spacecraft returned 2.6 billion bits of information, including 16,000 photos from the lander and 550 images from the rover, fifteen chemical analyses of rocks in the vicinity, and an entire archive of information on local environmental conditions."[4]

As amazing as the achievements of *Pathfinder* and *Sojourner* were and are, they pale before the promise of the Mars Airborne Geophysical Explorer, the *Kitty Hawk*, scheduled for December 17, 2003, the 100th anniversary of the Wright Brothers' historic flight at Kitty Hawk, North Carolina. A thirty-pound winged aircraft, developed by the Naval Research Laboratory, will be deployed from an entry vehicle descending through the Martian atmosphere. The *Kitty Hawk* will fly for three hours along a Martian valley, called *Valles Marineris*, at an altitude of 30,000 feet. *Kitty Hawk* will carry geophysical instruments, as well as still and video cameras. The spacecraft that unleashes *Kitty Hawk* will pass overhead, collect information from the aerial explorer, and relay it all back to earth.

Orville and Wilbur Wright might well have said: "Wow!"

Port Canaveral

Port Canaveral grew in fifty years from a small oil and shrimp port into one of the busiest cruise and commercial ports in the Western Hemisphere. The state legislature had authorized the Canaveral Port Authority in 1939, but World War II held up its actualization. The legislature finally approved it in 1945 and granted appropriations for needed construction the following year. The space program greatly increased its activity and use. Materials needed for space exploration could come by water.

That was one of the reasons the missile-makers moved from White Sands, New Mexico, to Huntsville, Alabama back in 1950. The Tennessee River offered a waterway for materials and finished rockets via the Ohio and Mississippi Rivers, through the Gulf and around the Keys to Cape Canaveral. So the port was to provide great service for the space program, and for the area in general.

The development of Port Canaveral is an interesting story. In 1955 a vessel loaded with newsprint and a petroleum tanker made the first Port Canaveral calls. Gradually, petroleum came to be one of the port's major imports. In 1958 Tropicana tanker-vessels began shipping refrigerated orange juice to New York. The port shipped bulk cement in the mid-1960s. Canaveral became a port of entry in 1961.

The first passenger ship in 1963 brought Lockheed employees to work at the Cape. Constructed in 1965, locks allowed access from the Atlantic Ocean to the intercoastal waterway. During 1966, the port's cargo tonnage reached the 1 million mark for the first time. Petroleum accounted for 93 percent of the cargo. As tonnage continued to increase through the 1970s, so did the variety of products. Scrap steel processed locally and fresh citrus went to northern Europe and Japan.

The first cruise ship came in 1971 and brought tourists to Disney World. Over the succeeding eleven years, eighty-nine cruise ships brought 54,000 passengers to visit central Florida attractions. In 1982, the Scandinavian World Cruises sent a boat, the Scandinavian Seas. In the first year, the ship handled 236,000 revenue passengers. Cruise terminals 2 and 3 were finished in May 1983 and began operations in 1985. At that time, Disney World joined in the port's marketing program.

All the while, the handling of cargo matched the growth of cruises. During the 1980s, citrus concentrate continued to be a big export. In 1982, the International Salt Company, later called the Cargo Salt Company, moved its operations from Port Tampa to Port Canaveral. Morton Salt Company also opened a salt processing plant at Port Canaveral in 1990 to evaporate seawater. By the year 2000 over a quarter of a million tons of salt were shipped through the port annually. In the early 1990s, the single strength orange juice came back after a thirty-year hiatus. Other primary cargoes at Port Canaveral—lumber, cement, and news-print—grew steadily.

The seafood industry thrives at Port Canaveral. When the first non-seafood cargoes were unloaded in 1955, only one cargo pier existed to handle dry and liquid cargoes. Over the years, Port Canaveral opened two liquid bulk facilities and nine dry cargo berths with 6,976 feet of berthing space, including two roll-on–roll-off ramps available for its customers. Future plans called for the construction of three additional cargo berths in the north cargo area. Port Canaveral south cargo piers 1, 2, 3, 4, and 5, provided over 3,200 feet of docks for frozen and perishable food shipments and other general cargoes. Covered dry freight storage capacity on port property totaled 750,000 square feet.

While many individuals were satisfied to ship tuna, others wanted to catch them, to enjoy the seashore, or to swim in the ocean. Responding to the wishes of the general public, the Port Authority wisely developed recreational activities on its land—camping, swimming, and fishing—and dedicated four acres as the Port's End Park.

In March 1987, Port Canaveral received its Foreign Trade Zone Alliance. By 1990, it received an estimated $413 million in revenues and had an economic impact of $1.4 billion on the local economy. In three years, the International Business Magazine rated Port Canaveral one of the top ten foreign trade ports. Growth continued through the 1990s. Fiscal year 2000 had a volume of 4.55 million tons, a 10 percent gain over the fiscal year 1999.

The Canaveral Port Authority worked closely with commercial cargo industries resident in the port to expand and diversify the cargo base for the future. A multiyear plan looked to increasing tonnage and meeting future market demands. The area welcomed private terminal and warehouse operators. Among these, Mid-Florida Freezer Warehouses, Incorporated, with 8.6 million cubic feet of space, claimed to have the largest privately held vessel-side facility in the South. The company offered more than 400,000 square feet of warehouses. Ambassador Services, Incorporated, with its headquarters at the port, offered a wide array of services: planning, scheduling, stevedoring on cruise ships, processing, fabricating equipment, operating rail terminals and warehouses.

The Foreign Trade Group functioned in a climate-controlled warehouse. It offered computerized inventory systems, management services, record storage, and value-added distribution services. U.S. Custom House brokers and freight forwarders were available onsite. The Integrated Distribution Services, Incorporated, situated within a few hundred yards of the docks in a climate-controlled warehouse, opened the first container freight station at the port in 1999. With foreign trade zone groups, it enjoyed U.S. Custom House broker services onsite.

Port Canaveral constructed a small container facility on the port's north side. This allowed the port to serve as a feeder to the much larger hubs in Charleston, Savannah, Miami, and Freeport in the Grand Bahamas. The building of this container facility completed the infrastructure necessary to fulfill all transportation needs of central Florida.

The NASA and Air Force leaders, who, shortly after World War II, chose the Cape as a center for launching vehicles into space, had planned a port of embarkation on the way to the moon, not a major commercial port. But, as a result of space exploration, the Cape also came to be a place where men sent cargo and people to many places in the world and, in turn, welcomed people and materials from other parts of the United States and elsewhere for space-related activities and, later, for wider purposes.

From Citrus Farms to Manufactories

Sociological studies have shown that even if the vast majority of people lived on pensions, at least 33 percent of any community would be involved in the service trades that provided for these people. In line with this study, had the space program chosen another locality, Brevard would have grown steadily in retail trade to provide services for tourists and newcomers seeking the warm South. But with the coming of the space effort, these figures on the growth of retail sales pale before the statistics on manufacturing.

When Jim Deese came to work at the Cape in the days of space beginnings, he had to drive to Orlando for almost every need. The sixties and seventies saw dramatic changes. The 1980s brought a high of 181 new manufacturing firms in the county, and the 1990s saw another 122 of those that listed the year of their establishment. The *Space Coast Manufacturers Register* for the end of the millennium described 831 firms with the addresses, names of general managers or contact persons, years of establishment, and areas of production.

In 1950, Harris Corporation in Melbourne, with forty employees, was the only manufacturing concern of any consequence in the county. By the year 2000, Harris Corporation employed 6,000. Three hundred other firms employed 32,305 workers in manufacturing and 11,357 in construction. The average wage in manufacturing was $39,722 and in construction $25,435.[5]

Harris shared electronic business with five other plants in North Brevard alone. By the year 2000, ten printing shops promised excellent letterheads. Seven firms built boats, thirteen manufactured metal products, and seven dealt in plastics. In all, there were seventy categories of manufacturing. No longer did one have to go to Tampa for uniforms, for wrought iron work, or for upholstering. One could even find a firm that made golf cleats and another that dealt in seashell products. A third specialized in T-shirts with clever or fancy symbols on the front. The athletic department at Astronaut High could get trophies for its star athletes right at home. The chemical business was thriving, too, with five chemical concerns. Eight dealt in concrete. If someone wanted a sign painted, six different firms could offer bids.

Some of the titles of the firms showed imagination. The Village Smithy, for instance, promised to manufacture and install ornamental wrought iron doors, railings, and window guards. One might see the name Catzban Diversified and think it was a misspelling of Cat Scan. But when he met Johnny Catzban, the owner of the firm, he would realize that the firm bore the family name. Sea Stuffe, Incorporated, dealt in seashell products, especially trinkets made of seashells.

Brevard also had five custom woodworking firms and three that dealt in marble. One firm made doors, another fixtures. Others specialized in rigging, air-conditioning, or cabinetmaking. Another dealt in ceramics and others in spraying equipment, fiberglass, wire, gadgets, tools, upholstery, cabinets, and frames. One firm coined its own category: "signage." Signs and Designs, Incorporated, manufactured plastic, wood, metal, and vinyl signage. No doubt Noah Webster's successors will eventually catch up with the pioneers who used the word down in Brevard, as they had done with local meanings of "interface" and "parameters."

Prior to 1950, social observers often spoke of four Floridas. In "Plantation Florida," west of Tallahassee, large landholdings with rich soil, reminiscent of Tara in *Gone with the Wind*, were well suited for cotton production. Around St. Augustine on the Atlantic coast, "Spanish Colonial Florida" clung to remnants of prerevolutionary days: impregnable Fort San Marcos; residences resonant of Iberia; the ancient shrine of *Nuestra Señora de la Leche*; a cathedral built at a later date but in colonial mission style; and a population whose ancestors came in the days when pirates still menaced the Spanish Main.

"Farming Florida" stood between these regions and stretched off to the south. Miles of sandy soil challenged isolated farmers who broke the sea of pines and palmettos to raise corn, cattle, and hogs. Marjorie Kinnan Rawlings's novel, *The Yearling*, and the heart-warming movie based on it, introduced these Floridians to the nation.

"Tourist Florida," areas along the Gulf and the Atlantic with elegant hotel accommodations, lured winter guests from the North and Middle West. Adjacent to them, countless communities began to welcome retirees as air-conditioning made possible comfortable year-round living in the South. The scattered citrus groves might merit a special category, as well as the cattle industry that followed the introduction of new breeds for tropical areas. We leave that to the sociologists.

But clearly, in the second half of the twentieth century, space exploration gave birth to a "Fifth Florida," an area of vital aerospace firms and their many subsidiaries.

Harvest of Lift-offs and Landings

The extent of space-based knowledge dazzles the imagination. Back in 1962, as mentioned earlier,[6] NASA Director James Webb set up a Technological Utilization Program to highlight the items of space technology used or usable in civilian economy. By 1973, the program pointed out 30,000 such uses with new ones rolling in at the rate of 2,000 a year.[7] NASA described these areas in various publications and set up seven dissemination centers to work directly with industry. In 1973 alone, NASA helped 2,000 companies and fielded 57,000 queries. New products ranged from quieter aircraft engines to small calculators.

Thanks to space explorations, mechanics had at hand cordless tools; an electromagnetic hammer—developed for use on the Saturn V at Marshall Space Center—that gained the attention of aircraft, ship, and automobile manufacturers; and the "O-ring shock absorber," developed by NASA and later employed by many states in their highway barrier systems. All people had available new fabrics, lightweight and heavily insulated for clothes and blankets. The handicapped now had a motorized wheel chair, activated by a sight switch, that gave more than 100,000 paraplegics greater mobility. This came from a device to permit astronauts to operate space controls when strong gravitational forces prevented movement of their arms. A voice-controlled wheel chair developed, based on tele-operator and robot technology. Non-handicapped seniors have their blessings, too. An EZ-UP lift helps them get out of their chairs.

From the scene of accidents, ambulance teams can transmit electrocardiograms to hospitals by radio and telephone. A Supercritical Air Mobility Pack (SCAMP) assists in the detection of breast cancer. A tracking transmitter, the Personal Location Beacon, sends distress signals.

The Earth Resources Technological Satellite provides information on crop growth; the use of grazing lands; the ecological effects of the meandering of the Gulf Stream off the east coast of the United States; the formation and location of icebergs; the precise area where herring feed at a given time; the location of major ore deposits; the occurrence of storm and tidal erosion on our coasts; and the extent of timber resources. It can give a clear picture of the extent of snow cover in the high Sierras, of pollution of lakes in Wisconsin, of infestation of crops in Iowa, and of land use in the clustered cities of the country.

Many other spin-offs followed manned space exploration. Teflon, a new coating sprayed on frying pans, became commonplace. In the area of food packaging, astronaut needs created food containers with pop-open lids, now commonplace. A private company developed them for manned launches. Space experiences brought about devices for preserving fruit and improving potato yields. A quipster at KSC called these new potatoes "Space Age Spuds." After eating, the astronauts cleaned their hands with Kim-Wipes. Soon non-astronauts used them at KSC and elsewhere.

Ribbed swimsuits enhanced speed. Golf balls gained greater velocity and carried farther, even when a duffer hit them. Sunglasses, called the "Blue Blockers," protected vision by thwarting blue, violet, and ultraviolet light that caused eye disorders, such as cataracts and senile macular degeneration. NASA also developed scratch-resistant lenses. Space-developed electroluminary panels were used for nightlights in many forms from wristwatches to exit signs in buildings.

Foam cushioning, initially marketed under the name "Temper Foam," became, according to an article in the magazine *Spinoff 1988*, "one of the most widely used spinoffs of NASA research."[8] It went into aircraft seat cushions, dental chairs, parlor furniture, football helmets, and chest protectors.

In conjunction with a water firm, WWI in Forestville, Maryland, NASA developed AquaSpace Industrial Filters. Before the space program, humans could cool rooms or cars. Now they can cool suits. Originally designed by NASA to provide cooling systems within astronaut space suits, a custom-made suit can circulate coolant through tubes to lower the body temperature and bring about dramatic improvement in patients with symptoms of multiple sclerosis and cerebral palsy.

To accommodate the tight confines of space travel, NASA had to replace bulky refrigerators with miniature thermoelectric components. Today portable coolers and warmers plug into the cigarette lighters of autos, boats, and other recreational vehicles. Space research led to water treatment systems used today in developing countries. To allow firefighters and others to work in intense heat, NASA developed a type of fire-resistant material called "Nomex." Racecar drivers purchased Nomex suits and were soon walking away from flaming cars.

On the other end of the temperature scale, NASA developed an ice-making machine. Called ICEMAT, it preserved the surface for a twenty-two-day-long run of an ice show in Dallas, even though the daytime exterior temperature ranged in the mid-90s.[9] Programs and studies at the spaceport led to advanced book preservation techniques and to new technologies for discovering and studying paintings by old masters that had been painted over at a later date. Aerial photographers found forgotten graves at Mims in North Brevard.

Among other space spin-offs were surveillance systems, microcomputer software, a computer reader for the blind, a microbe detector, and acoustic materials called SMART Products that lessen noise and air pollution. Many of the spin-offs greatly advanced medical efforts to treat heart diseases. Among such developments were pacemakers, angioplasty, cardiac imaging systems, an ear thermometer that worked instantaneously, a blood analyzer, and electrophoresis—a widely used method of determining the presence and amount of specific blood constituents that employs an electrical current to separate fluid components and prevents interference from other compounds in the solution.

NASA spawned several monitors. A wind measurement balloon for Apollo launches measured the wind for meteorological studies and predictions. The Jimsphere, a wind-measuring balloon four feet in diameter and made of lightweight radar-reflective materials, became a standard monitoring device at all missile ranges. It is now in commercial production. A NASA software package measures the concentration of target gas. It separates various gases in smokestacks and determines the amount of individual gases present within the stream for promoting compliance with sound emission standards.

NASA developed a self-righting life raft for astronauts to use while waiting to be picked up after ocean landings. This device has wide use on fishing boats, pleasure craft, and rescue vessels. It became standard rescue equipment for the Coast Guard and the Navy. A type of rubber developed for tires on the lunar rover and material developed for Viking lander parachute shroud lines was adapted for use in radial tires. With this material, tires maintain their flexibility in below-freezing temperatures. The shroud line material, five times stronger than steel, replaced steel belts and created a much more efficient tire for cold climates.

To train astronauts, especially those chosen to spend time in space, NASA developed a conditioner to counter the weakening of the astronauts' cardiovascular systems. This device is usable in other areas of physical training besides going into space. NASA made efforts in the channeling of solar energy. It also took part in improving water purification processes.

NASA software engineers developed a program that mathematically analyzes a design and predicts how it will withstand various stresses and strains. The NASA Structural Analysis Program or NASTRAP can be used in many non-aerospace applications—automobiles, machine tools, and hardware designs. A NASA procedure designed to sterilize water, with minimum power expenditure and little monitoring, on long-duration flights has found commercial use in swimming pools throughout the country.

Space shuttle tests of an optical lighting detection technique led to the development of personal lighting detection systems that began to be widely used by golfers, boaters, homeowners, and private pilots. The Jet Propulsion Laboratory developed an emergency response robot that could take the place of humans in performing hazardous tasks.

NASA developed new school bus designs that Navistar International uses in the production of 50 percent of school bus chassis. NASA also developed a new wing design used in business jets built by Gulf Stream. An energy storage system provides continuous power for critical electrical systems. New lubricants for railroad track manufacture as well as rust and corrosion prevention grew out of NASA's studies. An improved aircraft engine, such as the one on Boeing 777, stemmed from NASA-developed technological designs.

In short, the research, study, and experience of the people of KSC, alongside their affiliates and constituents, has driven development of many devices, great and small, that have improved the human condition.

9

REORIENTING OUR DREAMS

Changes at the Visitor Center

Over the years agencies of the federal government at the Cape had handed over to private concerns various aspects of the operation. Back in 1953, the Air Force awarded Pan American a contract for day-to-day operations that included setting up cafeterias and providing security. The Radio Corporation of America received a subcontract for technical aspects of range operations. In 1964, NASA moved on three contracts. In January, Ling-Temco-Vought took responsibility for photographic support, technical information, a field printing plant, and administrative automatic data processing. In February, TWA won a contract for supply, general maintenance, and utilities. In April, Bendix signed a contract for various services that included operating the crawler-transporter.

None of these contracts affected the average space worker. The 1995 turning over of the KSC Visitor Complex to a private firm that had to charge an entrance fee, even to members of the spaceport community, touched a nerve of space veterans who looked upon the original Visitor Center as their own. "The people's money built the center," they exclaimed. "It belongs to the people. Visitors should pay for bus tours and the purchase of souvenirs. But the displays should stay in the public domain." The story of the arrival of a busload of children from North Carolina, who lacked the entrance fee and were turned away, brought the change of procedure to the attention of the entire community. To the veteran space-worker, it was as if a novice college president had told alums that they were unwelcome on campus. NASA might wisely have made special arrangements for veteran space-workers.

Perhaps the announcement that the Delaware North Parks Services had more than $1.3 billion in revenues made the public uneasy. The nineties were a time of wealth-flaunting, much like the period of "conspicuous consumption" a century earlier. Reports of firms that registered excess profits while laying off workers and the financial manipulation of multimillionaire CEOs rankled the average American.

Delaware North could have made a better presentation of its reason for charging an entrance fee. The Visitor Complex was entirely self-supporting. It received no government funds. The percentage of visitors who purchased tickets for special events or bus rides or purchased souvenirs failed to provide sufficient funds to cover the routine expenses that running visitor programs entailed, let alone allow for necessary improvements.

Wisely, Delaware North Parks revamped its public relations department. The new director of that department, Pashen Black, set out to do what should have been done earlier. The company spent a large amount of money to enlarge the scope of the Visitor Center. These improvements came at the company's expense, not from taxpayers' dollars. A better feeling began to grow among the old-timers.

Within five years, and after a large loan from the Spaceport Florida Authority, the Visitor Complex saw an enriched facility. The number of visitors grew by 35 percent and averaged 2.8 million visitors a year.[1]

In its 1997 brochure, KSC welcomed tourists to the Visitor Complex. Parking for the family car was provided on a toll-free lot south of the complex offering easy access to Spaceport Central with its large number of fascinating and informative exhibits and the Space Theater. At the information counter one could obtain maps and the schedule of tours throughout the day. These helped visitors understand what they were seeing, from communications satellite displays to models of the huge launch facilities on the bus tour.

The Gallery of Space Flight to the north of Spaceport Central featured Mercury, Gemini, and Apollo spacecraft. These historic exhibits provided a fascinating study of the early days of the American manned space flight program. To the west of Spaceport Central, the visitor could stroll through the Rocket Garden that took the newcomer from the comparatively tiny Mercury Redstone that lifted the first American into space to the huge Saturn IB vehicle that launched the three-man Apollo crews into earth orbit. The Shuttle Plaza on the east side of the Visitor Complex contained a full-size model of a shuttle Orbiter. Visitors could enter it for a realistically detailed look at the cargo bay, the living quarters, and the cockpit. Nearby stood two giant solid rocket boosters, external tanks that provide the brute power needed to lift the space shuttle toward orbit.

The Galaxy Center, with its giant IMAX theaters, was located on the north side of the Visitor Complex. Movies were available, and the visitor checked at the information counter for time and theater. At IMAX, a complex camera system projected to a curved screen fifty feet high and seventy feet wide. The breathtaking imagery was supported by a six-track stereo sound system. One IMAX movie gave the visitor the closest possible simulation of the loud and fiery liftoff of the Space Shuttle. The other movie brought the visitor into the vast emptiness and sometimes eerie silences of space.

The NASA Art Gallery, a large collection of paintings by well known artists that depicted various aspects of the space story, covered the east end and the walls of the Galaxy Center. This unique collection was a "must-see" for fans of space art. Other exhibits stretched along the hallways. A huge memorial to those astronauts who died in the line of duty stood directly north of the building.

NASA invited professional educators to visit the Educators Resource Center located in the Center for Space Education in the northwest corner of the Visitor Complex. This installation offered extensive resources to aid the teacher in preparation of aerospace-related subjects, including a large number of free aerospace publications and 35mm slides, videotapes, and text data. Exploration Station, a "hands-on, minds-on" student science center, was another feature of the Center for Space Education. It contained two auditoriums—one devoted to lectures and science demonstrations, the other to science activities—and was open to everyone from 9:00 a.m. to 5:00 p.m.

The next stop, the newest visitors' attraction in the area, the Apollo-Saturn V Center, displayed numerous artifacts and exhibits from the Apollo years. It featured the Saturn V-Apollo that took the astronauts to the moon. The enormity of this 363-foot tall spacecraft astounded the average visitor. So rich was this treasure-house for the average visitor that many felt the need to stop, rest, and reflect lest they succumb to "astronautic indigestion." The first visit often proved overwhelming. As a result, KSC always counseled a full day's visit. During the day, a visit to one of the fast food shops and a relaxed lunch in the sun restored the stamina of many.

Without taking a bus tour, the visit seemed incomplete. On such a tour the newcomer could see the Vehicle Assembly Building, the space shuttle launch pads, the huge mobile launch platforms whereon the space shuttles were assembled, and the ponderous crawler-transporters that hauled a shuttle on its platform to the pad. As if this were not enough for one day, the bus tour also took the visitor to the New Frontiers Gallery and to two theaters. The Firing Room Theater and the Lunar Surface Theater immersed viewers in the triumphs and setbacks that met President Kennedy's challenge to land astronauts on the moon and bring them back safely to earth.

The separate tour of Cape Canaveral Air Station included a visit to the Air Force Space Museum. Here the visitor viewed versions, early and late, of the Delta, Atlas, and Titan vehicles that had been the backbone of the unmanned launch operations for NASA. These vehicles had made the names Cape Canaveral and "the Cape" recognized around the world. The Space Museum, set up around launch complex 26, displayed a Redstone rocket on the pad whence America's first satellite, Explorer I, rode a Redstone into orbit in 1958. Not far away stood launch complex 5, from whence astronaut Alan Shepard became the first American to go into space on May 15, 1961. Here, on the threshold of space, each visitor had the opportunity to gain a new appreciation of the past, present, and future of the space age.

"Privatizing" the Shuttle

At the beginning of the fiscal year 1997, NASA turned over responsibility for space shuttle Orbiters, flight and ground operations, and shuttle logistics to the United Space Alliance—a joint venture of Lockheed Martin Corporation, Bethesda, Maryland, and Boeing Company of Seattle. It took in about $1.4 billion in information technology-related revenue from NASA. Originally, the United Space Alliance was between Lockheed Martin and Rockwell International Incorporated, but Boeing became involved when it bought Rockwell's aerospace and defense business in December 1996, about two months after the space flight operations contract took effect. Later, Boeing purchased McDonnell-Douglas. The performance-based contract is a six-year, $7 billion pact that also has two-year options.

To earn its fee, the United Space Alliance had to meet goals in four areas. Forty percent of the fee was based on safety and quality standards and another 40 percent on meeting mission and schedule objectives.

Reducing costs accounted for 18 percent of the fee. The final 2 percent was a program plus fee for bringing additional work to the shuttle program. The contract covered about a third of NASA's $3.2 billion annual budget for space shuttle operations. As other responsibilities were transferred to the United Space Alliance, that percentage would double to two-thirds of NASA's shuttle budget. By the time the United Space Alliance had run the space shuttle operations for one year, it had rocketed to the number two spot on *Washington Technology*'s "1999 Top Hundred List of Federal Prime Contractors."[2]

Space experts believed that this indicated the way NASA would run its operations, such as the space station, in the future. Among them was John Logsdon of the Space Policy Institute at George Washington University in Washington who opined: "I think NASA is in an important transitional move, away from government control to full private-sector control of space operations."[3]

NASA administrator Daniel Goldin had initiated the privatization move as a way to reduce the overall cost of human space flight. Under the contract, NASA no longer supervised day-to-day operations, but remained responsible for safety and high-level management. The responsibilities of United Space Alliance included astronaut and flight controller training, space shuttle flight simulator operations, mission planning, flight design and analyses, space station operations and utilization, vehicle modifications, testing, and checklist operations, ocean retrieval of solid rocket boosters, and controlling the space shuttle logistics depot.

As the contract moved forward, the United Space Alliance would pick up more responsibilities, such as working on external fuel tanks, main engines, reusable solid rocket motors, and solid rocket boosters. Over a period of six years of the base contract, the United Space Alliance was supposed to reduce NASA's costs by $400 million.

In the first full year of operations, the alliance trimmed costs by about $240 million by streamlining support functions, consolidating operations of several contracts, and eliminating duplication among various NASA contractors. By December 1999, Alliance president Ross Turner could state, "NASA overall became leaner and more efficient."[4] Cost reduction held top priority. To reduce costs, the company cut the number of workers by five hundred. Four hundred Space Shuttle workers lost their jobs in Florida. To meet this upsetting situation, the United Space Alliance established a career transition center in Titusville. Most of the four hundred displaced workers found other work in Florida. The salaries of administrators remained at the same high level. A proportionate decrease on this level might have eased feelings of stress. Such an action might have revitalized the feeling of partnership and cooperation that traditionally marked the space center.

The alliance claimed that shuttle operations had improved under its guidance and contrasted figures from 1989 and 1990—when twenty-three in-flight malfunctions occurred per shuttle flight—with the first two years of the United Space Alliance, when there were only seven. Turnaround time improved eighty-seven days from the return of an Orbiter shuttle to its next launch. Another goal of the alliance was to bring commercial payloads to the shuttle. If this came about readily, the Space Shuttle would add extra shuttle flights. The United Space Alliance hoped to keep its record with the Space Shuttle and with operations of the space station.[5] On that, of course, the United Space Alliance will not be the only bidder. Several aerospace companies might form an international consortium similar to the United Space Alliance to bid on the space operations contract.

Amid the glowing press releases that have accompanied this move, one looks in vain for a record that it has been considered deeply along with possible methodological alternatives. While scientists approved wholeheartedly, social scientists wonder why in its privatization policy NASA made no effort to expand industrial democracy to match our cherished political democracy. With a Democratic president professing great concern for the average American and deploring the excessive concentration of wealth, one might have expected provisions in the privatizing agreement to ensure that the workers of these powerful firms, along with the general American public, become economically involved, following the example of the West German government when it moved out of ownership and control of *Volkswagen*.

Almost a century and a half has passed since political scientist John Stuart Mill, considered one of the most intelligent men of his time, gave the opinion that ownership of factories by the workers was the only way to prevent excessive concentration of wealth and power, was, in short, the only way for capitalism to save itself. In our own country, social analyst Joseph Husslein of St. Louis University, in his 1919 forward-looking book *Democratic Industry*, called for the ownership of industrial plants by the workers.[6]

The American taxpayer funded the space program in all its stages. Many might ask therefore whether or not this privatization served the American people in the best possible way.

Postage Stamps and Steady Plans

Just as two events in the spring of 1965, the dedication of the Vehicle Assembly Building and the inaugural of KSC Headquarters, had marked a milestone in space, so two meetings in the month of January, 2000, proved another major step in the story of the spaceport. On January 12, the United States Postal Service unveiled a new stamp commemorating the Space Shuttle. Two days later the first Florida Space Summit met at the KSC Visitor Complex.

The launching of the Space Shuttle had called the attention of the American people to the great achievements that preceded it. Before, spacecraft had been one-time vehicles. Now reusable Orbiters went into space, came back, and went off again. The success of the shuttle had been a momentous occasion in space history, generally unappreciated by the average American.

On the first day of the issue of the new stamp, Viki Brennan, Central Florida District Manager of the U.S. Postal Service, unveiled an enlarged commemorative stamp that showed the shuttle and the pad against the background of the American flag. As a part of the Postal Service's "Celebrate the Century" Series, it is one of fifteen stamps dedicated to poignant memories from the 1980s. Among these recollections, the Postal Service chose to commemorate the return of the Iran hostages, the fall of the Berlin Wall, and the dedication of the Vietnam Memorial. Lest one conclude that the eighties brought only grim memories, the Postal Service also recalled the football victories of Joe Montana, Jerry Rice, and the San Francisco Forty-Niners.

KSC Director Roy Bridges, who as pilot on the Space Shuttle *Challenger* flew a mission in 1985, spoke at the unveiling ceremony. "This is a particularly meaningful event to me," Bridges said. "It was almost a miracle to be able to take that first generation of reusable space vehicles up and return them safely home."[7] Perhaps many of those who heard Bridges say these words were a little surprised. By that time, most Americans had taken for granted that NASA succeeded in whatever it set itself to do.

When State Representative Dave Weldon and astronaut Richard Linnehan spoke, they paid special attention to groups of children in the audience. One group came from *Challenger* Elementary School in Port St. John, the other from the Mims Elementary School at the other end of the county. Among other remarks, Congressman Weldon prophesied, "Maybe ten to twelve years from now we'll be here to unveil an international space station stamp. At least I hope we will."[8]

Immediately after the unveiling, purchasers lined up to buy the new issue. Some wondered who chose the Space Shuttle and other images. "'Patrons' of the Postal Service," manager Brennan answered spontaneously.

Two days later, KSC saw one of the most significant meetings in space history. A motorcade, with lights flashing and sirens shrieking, approached KSC's Visitor Complex. Helicopters fanned the air overhead and alerted tourists to the fact that important individuals were about to descend on them. The day had the "feel" of an Apollo launch or of President Kennedy's first visit.

This first Florida Space Summit brought together a significant assembly of leaders from state and federal governments, from private industry, from NASA, and from the Air Force to discuss Florida's future in space launches. The conferees included NASA Administrator Daniel Goldin; Florida's Governor Jeb Bush; Florida Senators Robert Graham and Connie Mack; Representative David Weldon who, with Senator Graham, had organized the gathering; State University System Chancellor Adam Herbert; State Representative Randy Ball; KSC Director Roy Bridges; Forty-Fifth Space Wing Commander Brigadier General Donald Pettit; and the leaders of aerospace companies Boeing and United Space Alliance.

"This meeting is unprecedented," NASA Administrator Daniel Goldin said early in the two-and-a-half-hour session. "I have been administrator for eight years now, and I have never been invited to a meeting like this."⁹ Frankly, a meeting with that purpose surprises the outside observer who had long been confident that the future of space lay with Florida, not with any other American locale or any other country, including Russia. And yet, to his satisfaction, the leaders did not take that for granted. They gathered to make sure that Florida was the place.

After the political leaders had made their remarks and welcomed participants, Director Bridges and General Pettit outlined present activities and future plans of their respective areas. They stressed the importance of joint operations and gave a summary of the Spaceport Technology Center concept. The two men used a series of questions as starting points for exploring the needs and expectations of all concerned groups.

Lieutenant Governor Frank Brogan seemed to sum up the concern of others on the matter of Florida's growing competition for space launch business. "Our simple geography," Brogan said, "doesn't give us the complete advantage we once had. The pieces of the puzzle are there. Now it's a matter of interlocking them to create a picture of space in Florida that we want to see."¹⁰ He emphasized the state's efforts in securing funds for the space program.

Several speakers talked about the need for better coordination among the local, state, and national governments and with private companies involved in the space program. The lieutenant governor emphasized the state's commitment to obtaining funding for range upgrades and the space program in general. Several speakers thought that the state universities ought to be even a little more involved in the space program. Director Bridges said that one such opportunity will come about with the development of the Space Experiment Research and Processing Laboratory (SERPL), the latest of acronyms that abound at Kennedy Space Center. This facility, scheduled for construction near the KSC Visitor Complex, was to offer state-of-the-art equipment in a facility jointly operated by NASA and the scholars of the country. Plans called for the facility to become the processing center for NASA's life-science mission. Bridges told the gathering that SERPL was intended as the first stage in a 400–acre Commerce Park that could eventually turn the Visitor Complex into a twenty-four-hour facility.

University Chancellor Herbert said that the university system regards its future involvement with SERPL as a high priority and promised that the University of Florida would serve as lead institution for academic endeavors at the laboratory. NASA Administrator Goldin discussed the economic benefits Florida might derive from future efforts. He mentioned several possible developments in the immediate or distant future that would enhance KSC's value: robotic colonies throughout the solar system, commercial space travel, human colonies on the moon and other planets, and space stations in high orbit. He emphasized, however, that all of these ideas depended upon dramatic increases in safety and in reduction of the cost of space flights.

Earlier, Governor Bush had been satisfied to listen and learn. Eventually, he suggested that the summit become a regular event. Senator Graham pushed the idea of establishing a steering committee, representing all the key entities involved, as a means of monitoring progress. Lieutenant Governor Brogan reflected the views of many in his final remarks: "There are a lot of things we can do to position ourselves to take advantage of the brain trust here. Far too many people just take space for granted because it has been here, it is here, and they think it always will be on the Space Coast. We don't want to just hang on to the space program. We want it to grow in Florida."[11]

Few seem to share Lieutenant Governor Brogan's concern. In the last half of the twentieth century, the final decades of the second millennium, space exploration had changed the face of Florida from a citrus-growing and tourist-welcoming area to a major manufacturing state. It had made the name of Cape Canaveral a byword for vision and achievement all over the world, and improved life on earth in countless ways.

It was Florida's destiny. Geography chose the state's east coast for the spaceport that united planets with the same certainty that it chose the Isthmus of Panama for the ocean-uniting canal.

I first heard of NASA's request for a senior professor of history to work on the story of the Apollo launches from a colleague of mine at Saint Louis University, Dr. Charles Fleener. Several years before, he had finished his work in Latin American history at the University of Florida under the direction of Dr. David Bushnell, a member of the National Committee on Space History. Dr. Fleener learned of his former director's search for a history professor and writer to head a two-man team on the story of Apollo. Dr. Fleener asked if I was interested.

While few professors considered moving anywhere but to Harvard or Stanford at that time, the campus demonstrations and classroom confrontations of the turbulent sixties were taking away the ordinary gratification of college teaching. Ready for a working sabbatical of a year, I was less enthusiastic about a two-year assignment.

Nonetheless, I agreed to fly to Gainesville as a guest of the university and meet Dr. Bushnell and Charles Benson, a recent Doctor of History, member of Phi Beta Kappa, and an all-star football player, who was to be the associate historian. We enjoyed instant rapport. Later, on campus, I met three other faculty men who helped: Dr. John Mahon, the History Department Chair; Dr. Michael Gannon, the leading historian of early Florida; and Dr. Bernard Leadon, professor of aeronautical engineering. All at Gainesville proved most helpful, as did Dr. Eugene Emme, space historian at NASA headquarters in Washington and his associate, Frank W. Anderson, Jr.

Dr. Benson and I drove to the Space Center and met with James Russo, head of the division at KSC where we'd begin working in January 1972. I arranged for lodging at St. Theresa's Rectory in Titusville, a twelve to fifteen-minute drive to Kennedy Space Center.

My co-worker, Dr. Charles Benson, and I went to KSC in January 1972. To move from the more relaxed atmosphere of university life to the intense concentration of the Space Center took adjustment. The greatest difficulty lay in the type of contract we were to work under—a "level of effort" contract whereby we put in time each day, from 8:30 to 4:30. Writers do not work on someone else's schedule but develop their own.

On weekends, I assisted Fathers Michael Hanrahan and Declan Jordan with church duties at St. Theresa's. The people of Titusville held both young natives of Ireland in high esteem. It was obvious why. Like St. Francis, the gentle and kind pastor Father Hanrahan preached best by example. Father Jordan, a vigorous young priest and like the pastor a joy to know and work with, was especially successful in youth ministry. A parishioner, Major Walter Wagner, agreed to drive me from Titusville to the Space Center every morning until I was able to rent a car.

In those early months, people in all areas of the spaceport, and especially the parishioners of St. Theresa's Parish in Titusville, were anxious to show me what they did at KSC. As a result, I was able to watch the stacking of Apollo 17 in the VAB, to view its roll-out at close range, and, on the occasion of bad weather, its return to the assembly building. Even more memorable, I enjoyed a simulated landing on the moon in a command module.

My previous writing routine called for steady work in the morning, a period of relaxation, a swim or a game of tennis some time in the afternoon, and then rewriting in the evening. This routine clashed with the KSC schedule. The morning hours were fine, the afternoon difficult, and the evening bad. After dinner in the evening, when I ordinarily did rewriting, I was away from the Space Center at St. Theresa's Rectory in Titusville. Gradually, I adjusted to the new schedule.

I could pass on to my colleague tips on getting down to write in the morning or counteracting the fed-up feeling that usually lasted for two weeks. I had learned the tricks of the trade from Father Raymond Corrigan and Clarence McAuliffe at my alma mater, from Tere Rios and August Derleth at the Rhinelander School of the Arts, from Milton Lomask and Patricia McGeer at the Georgetown Writers' Week, and from Bernadette Hoyle at the Tar Heel Writers Conference.

To begin the day, I rewrote the last page of material I had written the previous day. When sated with one aspect of the story, I learned to turn to a personality who grabbed attention. In earlier works, two such stood out: a magazine publisher who burned his bridges in front of him by firing editors before a qualified successor was at hand; and a school administrator who ignored those who supported him and favored the two who opposed his appointment. Writing about them always put me back on the creative track without a waste of time. Gunter Wendt at KSC overshadowed these two. When surfeited with complicated problems of engineering, I found that learning and writing about Gunter readily recharged the batteries.

In interviews, Dr. Benson and I came from different vantages and with interest in different aspects of the Apollo story. We interviewed widely, occasionally by phone to Huntsville or Houston. Some of those I interviewed seemed to feel more at ease because I was a clergyman as well as a trained historian and writer. Others felt uneasy for precisely that same reason. Sensing that, I asked Dr. Benson to talk with those individuals.

Some may have wondered at the time why the University of Florida hired a clergyman as senior member of the Apollo team. It was not as a priest but as a historian that Dr. Bushnell chose me, a full professor at Saint Louis University, to join the faculty in Gainesville for the duration of the project. It was not because of his doctorate in canon law that history remembers Niklas Koppernigk (Copernicus in Latin), a Polish priest, but because of his scientific studies and speculations that laid the foundations for modern astronomical science. Further, not as a priest, but as an astronomer, a century later, Giovanni Ricciolo, who belonged to the same religious brotherhood, the Jesuit Order, as I do, gave the names "Sea of Tranquility" and "Ocean of Storms" to craters on the moon.

We two historians divided the work so that Dr. Benson concentrated on the scientific areas, while I considered those areas dealing with the social, economic, and political aspects of Apollo. We worked together on the launches. It proved a quite satisfactory arrangement. My co-worker's father, Charles Benson, Sr., a retired journalist from the staff of the

St. Louis Globe-Democrat, was always ready to go over the writing. As a result, instead of distinct styles of writing, a consistent flow marked the book, *Moonport*, that came from our efforts.

Besides our Apollo history project, my associate Dr. Benson and I had mutual interests: world history, the St. Louis Cardinals baseball team—then getting second wind after the glory years of the sixties—and the Miami Dolphins, the football champions at the time. We also shared interest in one participation sport—tennis. Dr. Benson was almost as good at tennis as he had been as an all-conference quarterback at Davidson during his college days. Occasionally, in the lunch period, we played a set at the recreation center on the Banana River just south of headquarters. He resided in Daytona Beach and came daily with carpoolers to the Cape.

In Titusville, many groups invited me to give invocations or talks on history. Scout leaders asked me to serve on examining panels for prospective Eagle Scouts. The first project we approved called for a litter-free Titusville. The prospective Eagle Scout found discarded metal bins, painted them green, and placed them at strategic sites around the town. The Dean at Brevard Community College asked me to teach a survey of Western civilization, a course I had taught at Regis College in Denver and at Saint Louis University over many semesters. My students cut across the Titusville adult community. Jean Charron, a native of New Hampshire, secretary at St. Theresa's Church, and Newton Gregg, an engineer at KSC and a graduate of Southern Methodist University, typified the adults who attended. They wanted to enlarge their world vision.

The first launch that I witnessed, Apollo 16, on April 17, 1972, sent astronauts John D. Young, Thomas K. Mattingly, and Charles M. Duke into space. What I most remember, standing alongside the VAB and witnessing the launch, was its beauty. In reminded me of the sparklers we held in our hands as youngsters at the Fourth of July fireworks. Further, the vibration bouncing off the VAB almost knocked me down.

Minutes later, while I was talking to Hal O'Connor, director of the Merritt Island National Wildlife Refuge, a photographer came over and asked how the launch was. When we said it was beautiful, he explained, "At every launch, the brass has me photograph the VIPs and their reactions to the launch. So I have seen VIPs reacting to the launch a dozen times, but I have never seen one myself." Then he said, "Why don't you two let me take your picture?" As a result, Hal O'Connor and I ended up in the VIP file, along with Vice President Nixon and other dignitaries.

At that time, we met two of those dignitaries—the all-star quarterback of the Baltimore Colts, Johnny Unitas, and astronaut Gene Cernan, who had flown on Gemini 9 and circled the moon on Apollo 9, but had not landed. Both Unitas and Cernan were wiry men, lean rather than heavy-set, men who looked as if they were ready for whatever lay ahead. Cernan suggested that I come to Houston to meet the other Apollo astronauts. That opportunity never came to be.

Cernan had his opportunity to walk on the moon on Apollo 17 with Dr. Jack Harrison Schmitt, the first scientist on Apollo. One of the most sensational of all launches, it came shortly after midnight on December 7, 1972. When the motors roared, the light was so intense that the whole Indian River matched the white of Pike's Peak after a fresh snowfall. Fourteen days after returning from space, Dr. Schmitt talked at the Space Center. There we met. When he heard I was from St. Louis, he said that he had given the name of St. Louis in two forms to places on the moon. One he called "Spirit" after Lindbergh's plane, *The Spirit of St. Louis*; the other he called "San Luis Rey," after his favorite novel of college years, *The Bridge of San Luis Rey* by Thornton Wilder.

Even though, before signing up with the history, I preferred a one-year rather than a two-year stay, my second year in Florida was the happiest of my life up to that time. The book was on the flow, the spirit was high. The teamwork at the spaceport carried over into other aspects of life. In succeeding with the Apollo launches, the people in space exploration had come to know that everyone is important. That feeling of mutual appreciation carried over into community organizations, citizens' clubs, church groups, and PTAs. Everyone at the spaceport was important. Leaving the space area made them no less important.

St. Theresa's Parish held a farewell party for me in January 1974. It was one of the happiest evenings in my life. I knew I would come back. And I have, every year since then, revisiting KSC and reminiscing with friends about the glory days of moon exploration.

During the year 2000, Meredith Morris-Babb, editor-in-chief at the University Press of Florida, invited me to undertake a study of the impact of the space program on the development of the state for the Florida History and Culture Series. I was happy to accept her gracious invitation and to return to my former home, overlooking the Indian River and the launch pads to space.

Chapter 1. Testing the Territory

1. *Florida Today*, July 23, 2000, p. 85.
2. Wallace Dal Santo, interview by author, May 15, 2000, in author's file.
3. Dal Santo, interview.
4. G. A. Tokaty, "Soviet Rocket Technology," in *Spaceflight* 5, no. 2 (March 1963).
5. "He builds for space explorers," reprint from *Newsweek*, Aug. 16, 1965; see also Richard J. Whalen, "Banshee, Demon, Voodoo, Phantom and—Bingo!" reprint from *Fortune* 69, no. 5 (Nov. 1964).
6. Joseph Szofran, interview by author, tape recording, May 12, 2000, in author's file.
7. *Census 1960*, Florida, Table 86, pp. 11–251.

Chapter 2. Answering the Call of President Kennedy

1. Roger E. Bilstein, *Stages to Saturn: A Technological History of the Apollo/ Saturn Launch Vehicles* (Washington, D.C.: NASA, 1980), pp. 62–63.
2. Milton Rosen to Hugh Dryden and James Webb, Aug. 18, 1961. Quoted in Benson and Faherty, *Moonport: A History of Apollo Launch Facilities and Operations* (Washington, D.C.: NASA, 1978), p. 93.
3. Raul Ernesto Reyes, interview by author, May 17, 2000, in author's file.
4. Colonel Asa Gibbs, interview by James Covington, August 7, 1969, in Covington's file.
5. Courtney Brooks, James Grimwood, and Loyd Swenson, Jr., *Chariots for Apollo: A History of Manned Lunar Spacecraft* (Washington: NASA, 1979).
6. James V. Covington, "History of Apollo Launch and Support Facilities," unpublished, in KSC archives, pp. iv-26, iv-23.
7. Fred Renaud, interview by author, April 8, 1973.
8. *Spaceport News* 3, no. 17 (Sept. 3, 1964): 1, 3.
9. Tom D. Crouch, *Aiming for the Stars: The Dreamers and Doers of the Space Age* (Washington: Smithsonian Institution Press, 1999), p. 183.
10. *Spaceport News* 2, no. 12 (June 20, 1963): 1.

11. Hugo Young, Brian Silcock, and Peter Dunn, *Journey to Tranquility* (Garden City, N.Y.: Doubleday & Co., 1970), p. 158.

12. Ibid.

13. Robert Rosholt, *An Administrative History of NASA, 1958–63* (Washington: NASA, 1966), pp. 288–89.

14. *Washington Post*, Sept. 21, 1963, p. A-10.

15. Thomas' letter and the president's reply appear in Senate Committee on Appropriations, *Hearings*: Independent Offices on Appropriations, 1964, 88th Congress, 1st Sess., pt. 2, pp. 1616–18.

16. *Fortune* 68, no. 5 (Nov. 1963): 125–29, 270, 274, 280.

17. Quoted in ibid., p. 270.

18. Ibid., p. 280.

19. Congressman Olin E. Teague, "Remarks," May 1968; quoted in U.S. House of Representatives, *For the Benefit of All Mankind: A Survey of the Practical Returns from Space Investment*, Report No. 92-748, 92nd Cong., 1st Sess., Dec. 14, 1971, pp. 4–5.

20. Glenn T. Seaborg, Public Statement, quoted in ibid., p. 5.

21. Frank W. Anderson, Jr., *Orders of Magnitude: A History of the NACA and NASA, 1915–1976* (Washington, D.C.: NASA, 1976), p. 36.

22. Ibid., pp. 37–38.

23. Quoted in *For the Benefit of All Mankind*, p. 3.

Chapter 3. Moving Ahead with LBJ

1. Quoted in *For the Benefit of All Mankind*, p. 3.

2. I was one of those students at Saint Louis University.

3. *Moonport*, 272, 326–28.

4. *Spaceport News*, passim.

5. *Census of Population 1960*, Brevard County, vol. 1, part 11, table 27, p. 90.

6. *Florida Power and Light Company Report*, May 1963. Copy in Kennedy Space Center Library (hereafter KSC), Archives Department.

7. *Residential Survey*, May 1964. Copy KSC Library, Archives Department.

8. *1970 Census of Population*, Brevard County, vol. 1, part 11, table 10, p. 20, has comparative studies for 1960 and 1970 and percentages of increase.

9. Ibid., table 119, Social Characteristics of Counties, p. 475.

10. Ibid., table 104, Employment Characteristics for Places of 10,000 to 50,000, pp. 378–84.

11. Ibid., table 124.

12. Ibid., table 10, Population of County Subdivisions, 1960 and 1970, p. 20.

13. Ibid., table 120, p. 481.

14. Newton Gregg, Design Engineer, interview by author, May 12, 2000.

Chapter 4. Learning from Success and Failure

1. Robert Bielling, interview by author, tape recording, October 3, 2000.
2. Joseph Szofran, interview by author, tape recording, May 26, 2000, in author's file.
3. Dan Venverloh, McDonnell Engineer, interview by author, November 12, 2000.
4. Dan Venverloh, interview.
5. Joseph Szofran, interview.
6. Nordby to Faherty, July 10, 2000, in author's file.
7. Quoted in *Moonport*, p. 389.
8. House Subcommittee on NASA Oversights, of the Committee on Science and Aeronautics: Investigation into Apollo 204 Accident: Hearings: 90th Cong., 1st Sess., pp. 390–91.
9. *Moonport*, p. 444.
10. Ibid., p. 457.
11. Ibid., p. 457.
12. Quoted in ibid., p. 477.
13. José González, interview by author at Kennedy Space Center, March 19, 1972.
14. *For the Benefit of All Mankind*, p. 1.
15. Congressman Joseph E. Karth, address before the National Space Club on May 17, 1968, quoted in ibid., p. 5.
16. Quoted in ibid., pp. 6–7.
17. *Christian Science Monitor*, Aug. 29, 1969, p. 6.
18. Haynes Johnson, in the *Washington Post*, April 9, 1970, quoted in ibid., p. 6.
19. Ibid., p. 11.
20. Ibid., p. 28.
21. *Florida Today*, March 16, 1972.
22. Ibid.
23. *Spaceport News* 12, no. 20 (Oct. 4, 1973): p. 11.
24. Gordon Harris interview, Feb. 2, 1973.

Chapter 5. Getting a "Second Wind"

1. Robert Bielling, interview by author, tape recording, May 21, 2000.
2. *Time*, July 4, 1969, p. 38.
3. Kurt Debus, quoted in *Moonport*, p. 315. Astronauts, too, felt family tensions. In an interview, astronaut Charles Duke, who went on *Apollo 16* to the moon, tells the strain his concentration brought on his family relationships. Fortunately, his wife brought their married life into proper focus (see Tom Neven, "The Work of His Hands," *Focus on the Family*, March 2001, pp. 7–8).

4. Dr. Ronald C. Erbs, interview by author, July 17, 1974, quoted in *Moonport*, pp. 315–16.

5. Ron Nazaro, interview by author, December 10, 1973, quoted in ibid., p. 316.

6. *Census 1970*, Florida, passim.

7. United States Census Bureau, 1954 *Census of Manufactures*, Vol. 3, Area Statistics; 1954 *Census of Business*, Vol. 2, Retail Trade, table 104.

8. *First Research Corporation*, table 22, p. 111 (hereafter *FRC*), in KSC Library, Archives Department.

9. U.S. Census Bureau, *Census of Business*, 1972.

10. *FRC*, Table 5.

11. Bilstein, appendix E, p. 424 ff.

12. W. David Compton and Charles D. Benson, *Living and Working in Space: A History of Skylab* (Washington: NASA, 1983), pp. 236–37.

13. Ibid., p. 341. The unused Skylab II is in the National Air and Space Museum in Washington, D.C.

14. Bilstein, p. 401.

15. Compton and Benson, p. 372.

16. Tom Crouch, p. 292. Much of the material in this chapter leans heavily on Crouch's book.

17. Quoted in *Moonport*, p. 527.

18. Eli Ginzberg and Associates, *Economic Impact of Large Public Programs: The NASA Experience* (Salt Lake: Olympus, 1976).

19. Ibid., p. 10.

Chapter 6. Welcoming a "Benign Boomerang"

1. Homer E. Newell, *Beyond the Atmosphere: Early Years of Space Science* (Washington, D.C.: NASA, 1980), p. 411.

2. Ibid., 388–89.

3. J. P. Kump, "Kennedy Space Center's Jones Is Described as an 'Engineer's Engineer,'" *Stargazer* 11, no. 18 (Oct. 6, 1994): p. 2.

4. Quoted in Kump, p. 2.

5. Ibid.

Chapter 7. Appraising the Home Front

1. *St. Louis Post-Dispatch*, Nov. 20, 2000, p. 11.

2. *Spaceport News* 39, no. 2. (Jan. 28, 2000): p. 4.

3. *1990 Census of Population*, Florida, General Characteristics of Persons, Households and Families, section 2, table 76.

4. Ralph Hall, telephone interview by author, May 12, 2000.

5. Vern Jansen, interview by author, May 11, 2000.

6. O'Donnell to Faherty, Colorado Springs, October 25, 2000. Letter in author's file.

Chapter 8. Looking Beyond the Far Horizons

1. Crouch, p. 289.
2. Patrick Moore, *Mission to Planets* (New York: W. W. Norton, 1990), p. 44.
3. Crouch, p. 290.
4. Ibid., p. 294.
5. *Space Coast Manufacturing Register*, Rockledge, 2000, p. 32.
6. See chapter 2, p. 41.
7. 2000 Spinoff, NASA; see also NASA Spinoffs, "Bringing Space Down to Earth," a packet of 128 three-by-five cards. Each gives the technology and commercial use of a particular spinoff.
8. *Spinoff 1988*, passim.
9. NASA Spinoffs, "Bringing Space Down to Earth."

Chapter 9. Reorienting Our Dreams

1. Pashen Black to Janice Taylor, KSC, Sept. 18, 2000. Copy in the author's file.
2. *Washington Technology*, "1999 Top 100 List."
3. Quoted in ibid.
4. United Space Alliance Update, December 1999, issue 24, p. 3.
5. Ibid., p. 1.
6. Joseph Husslein, *Democratic Industry: A Practical Study in Social History* (New York: P. J. Kennedy and Sons, 1919), pp. 349, 356.
7. *Spaceport News* 39, no. 2 (Jan. 28, 2000): p. 2.
8. Ibid.
9. Ibid., p. 1.
10. Ibid., p. 5.
11. Ibid.

RECOMMENDED READING

Professor Michael Gannon of the University of Florida wrote *Florida: A Short History* (Gainesville: 1993) and edited *The New History of Florida* (Gainesville: 1996) that tells the state's history in twenty-two chapters. The editor himself contributed the second chapter, "First European Contacts." For other chapters, he called on specialists such as John K. Mahon and Gary Mormino.

Charlton W. Tebeau's *History of Florida* (Coral Gables: 1971) is a well regarded general history that focuses on economic and political development. Rembert W. Patrick wrote an easy-to-read short history of the state called *Florida Under Five Flags* (Gainesville: 1967).

David A. Bice edited *A Panorama of Florida* (Charleston, W. Va.: 1982) that contains much valuable information on history, industry, transportation, and education, with photographs, maps, and charts. He devotes slightly over two pages in an eleven-page section on "Exploration" to the space shuttle.

In *Some Kind of Paradise: A Chronicle of Man and Land in Florida* (Gainesville: 1998), Mark Derr gives a paragraph to Cape Canaveral and Brevard County and, in passing, mentions the Kennedy Space Center.

In the field of space exploration, succinct annotations make Katherine Murphy Dickson's *History of Aeronautics and Astronautics: A Preliminary Bibliography* (Washington: 1967) especially valuable. An excellent early work, *The History of Rocket Technology: Essays on Research, Development, and Utility* (Detroit: 1964), edited by Eugene M. Emme, head of the NASA historians in the late 1960s and early 1970s, carries treatises by historians and space experts such as Wernher von Braun.

Several NASA historical monographs deal with early space programs and NASA activities. Among them, Constance Green and Milton Lomask's *Vanguard: A History* (Washington: 1971) is an excellent account of what was supposed to be our first orbital satellite. Robert L. Rosholt's *An Administrative History of NASA, 1958–1963* (Washington: 1966) is a valuable discussion of NASA's structure and procedure in its earliest years. Loyd S. Swenson, James M. Grimwood, and Charles C. Alexander's *This New Ocean: A History of Project Mercury* (Washington: 1998) clearly meets the promise of its title. James Grimwood teamed with Barton C. Hacker on the official history of the two-man launches, *On the Shoulders of Titans: A History of Project Gemini* (Washington: 1977).

Edgar M. Cortright edited *Apollo Expeditions to the Moon* (Washington: 1975), a colorfully illustrated summary of the Apollo-Saturn program written by NASA astronauts and executives, among them von Braun who wrote the essay on Saturn.

In *Moonport: A History of Apollo Launch Facilities and Operations* (Washington: 1978), historians Dr. Charles D. Benson and I look at the Apollo launch story from outsiders' vantage points. *Moonport* is part of a trilogy including Roger E. Bilstein's thorough and readable *Stages to Saturn: A Technological History of the Apollo/Saturn Launch Vehicles* (Washington: 1996) that details the complex development of the boosters, and *Chariots for Apollo: A History of Manned Lunar Spacecraft* (Washington: 1979), by Courtney G. Brooks, James M. Grimwood, and James Swenson, Jr. Based on research and over 300 interviews, this well-written work covers the story of Apollo spacecraft through July 1969.

Roger E. Bilstein also has written other important space books: *The American Aerospace Industry: From Workshop to Global Enterprise* (New York: 1996), an outstanding overview of the history of this critical industry that gives due attention to the space flight aspects of its development; *Flight in America: From the Wrights to the Astronauts* (Baltimore: 2001), a superb synthesis of the origins and development of aerospace activities in America; and *Orders of Magnitude: A History of the NACA and NASA, 1915–1990* (Washington: 1989), a fine non-scholarly history of the National Aeronautics and Space Administration, and its predecessor, the National Advisory Committee for Aeronautics. Bilstein's *Orders of Magnitude* updates and extends a work of the same name (Washington: 1976) that Frank W. Anderson, Jr. published thirteen years earlier.

Two books by flight directors view space exploration from a unique perspective. Christopher C. Kraft and James L. Schefter wrote *Flight: My Life in Mission Control* (New York: 2001), a memoir detailing Kraft's tenure as NASA's flight director for Mercury-Gemini-Apollo and as Johnson Space Center director. Gene Kranz's *Failure Is Not an Option: Mission Control from Mercury to Apollo 13 and Beyond* (New York: 2000) tells the story of the flight director at the Johnson Space Center, to whom the movie *Apollo 13* gave proper recognition as a national hero.

Roger D. Launius has both authored and coauthored memorable books. *Frontiers of Space Exploration* (Westport, Conn.: 1998) presents documents and biographies of participants in space exploration. With coauthor, Bertram Ulrich, in *NASA and the Exploration of Space* (New York: 1998), Launius gives a basic history of NASA, illustrated with works from the NASA art program. With John Logsdon and Robert W. Smith, he coedited a collection of essays: *Reconsidering Sputnik: Forty Years Since the Soviet Satellite* (Amsterdam: 2000). In *Imagining Space: Achievements, Predictions, Possibilities: 1950–2050* with Howard E. McCurdy (San Francisco: 2001), Launius sweeps from past to future; and again with McCurdy in *Spaceflight and the Myth of Presidential Leadership* (Urbana: 1997), he published essays on the six presidents and presidential influence involved in shaping space policy.

Howard E. McCurdy has written several books of importance: *Faster, Better, Cheaper: Low-Cost Innovation in the U.S. Space Program* (Baltimore: 2001), a study of the history of NASA's efforts to reform itself in the 1990s; *Inside NASA: High Technology and Organizational Change in the U.S. Space Program* (Baltimore: 1993), a discussion of NASA's development from inception to the 1990s, using extensive interviews with key personnel as well as documentary sources; *Space and the American Imagination* (Washington: 1999), an analysis of the relationship between popular culture and public policy; and *The Space Station Decision: Incremental Politics and Technological Choice* (Baltimore: 1990), a fine study of the political process that led to the presidential decision in 1984 to develop an orbital space station.

A look at the moon launches from an astronaut's vantage, Michael Collins' *Carrying the Fire: An Astronaut's Journeys* (New York: 1974), won wide acclaim. Written many years later, Eugene Cernan's *The Last Man on the Moon: Astronaut Eugene Cernan and America's Race in*

Space (New York: 1999) gives new insights on NASA's efforts to put a man on the moon.

Charles D. Benson, my coauthor on *Moonport*, joined with W. David Compton to write *Living and Working in Space: A History of Skylab* (Washington: 1983). R. Cargill Hall wrote *Lunar Impact: A History of Project Ranger* (Washington: 1977), the mission that sent robotic probes to the moon in the late 1950s and early 1960s. Patrick Moore's *Mission to the Planets* (London: 1990) illustrates the story of man's exploration of the solar system: the Moon, Venus, Mercury, Mars, Jupiter, Saturn, Uranus, and Neptune. Tim Furniss reports on the shuttles from *Columbia* 1 in 1981 to Mission 22 in *Space Shuttle Log* (London: 1986). The extensively illustrated book includes photographs of and basic data on the shuttle crews up to that time.

Tom D. Crouch's *Aiming for the Stars: The Dreamers and Doers of the Space Age* (Washington: 1999), a publication of the Smithsonian Institution, sets out to tell "the story of the space age through the experience of those who were there"—from Johannes Kepler in early seventeenth-century Wurttemberg to John Glenn's return to space in 1998. Crouch's story of the Soviet's development of rockets fills a void in most American accounts of space exploration. This is a splendid account of the entire story.

In another publication of the Smithsonian Institution, Lillian D. Kozloski, has written the only serious study of space suits, *U.S. Space Gear: Outfitting the Astronaut* (Washington: 1993).

William Sheehan's *The Planet Mars: A History of Observation and Discovery* (Tucson: 1996), tells how humans have acquired knowledge about the Red Planet from antiquity to the present. It concentrates on the work of earth-based astronomers, but also includes succinct narratives of the Mariner 4 voyage and the Viking project of the 1970s.

Page numbers in boldface indicate photographs.

William Barnaby Faherty was professor emeritus of history at St. Louis University and director of the Museum of the Western Jesuit Missions in Hazelwood, Missouri.

The Florida History and Culture Series

Edited by Raymond Arsenault and Gary R. Mormino

Al Burt's Florida: Snowbirds, Sand Castles, and Self-Rising Crackers, by Al Burt (1997)

Black Miami in the Twentieth Century, by Marvin Dunn (1997; first paperback edition, 2016)

Gladesmen: Gator Hunters, Moonshiners, and Skiffers, by Glen Simmons and Laura Ogden (1998)

"Come to My Sunland": Letters of Julia Daniels Moseley from the Florida Frontier, 1882–1886, edited by Julia Winifred Moseley and Betty Powers Crislip (1998; first paperback edition, 2020)

The Enduring Seminoles: From Alligator Wrestling to Ecotourism, by Patsy West (1998)

Government in the Sunshine State: Florida Since Statehood, by David R. Colburn and Lance deHaven-Smith (1999)

The Everglades: An Environmental History, by David McCally (1999; first paperback edition, 2000)

Beechers, Stowes, and Yankee Strangers: The Transformation of Florida, by John T. Foster Jr. and Sarah Whitmer Foster (1999)

The Tropic of Cracker, by Al Burt (1999; first paperback edition, 2009)

Balancing Evils Judiciously: The Proslavery Writings of Zephaniah Kingsley, edited and annotated by Daniel W. Stowell (2000)

Hitler's Soldiers in the Sunshine State: German POWs in Florida, by Robert D. Billinger Jr. (2000; first paperback edition, 2009)

Cassadaga: The South's Oldest Spiritualist Community, edited by John J. Guthrie Jr., Phillip Charles Lucas, and Gary Monroe (2000)

Claude Pepper and Ed Ball: Politics, Purpose, and Power, by Tracy E. Danese (2000)

Pensacola during the Civil War: A Thorn in the Side of the Confederacy, by George F. Pearce (2000; first paperback edition, 2008)

Castles in the Sand: The Life and Times of Carl Graham Fisher, by Mark S. Foster (2000; first paperback edition, 2023)

Miami, U.S.A., by Helen Muir (2000)

Politics and Growth in Twentieth-Century Tampa, by Robert Kerstein (2001)

The Invisible Empire: The Ku Klux Klan in Florida, by Michael Newton (2001)

The Wide Brim: Early Poems and Ponderings of Marjory Stoneman Douglas, edited by Jack E. Davis (2002)

The Architecture of Leisure: The Florida Resort Hotels of Henry Flagler and Henry Plant, by Susan R. Braden (2002)

Florida's Space Coast: The Impact of NASA on the Sunshine State, by William Barnaby Faherty, S.J. (2002; first paperback edition, 2024)

In the Eye of Hurricane Andrew, by Eugene F. Provenzo Jr. and Asterie Baker Provenzo (2002)

Florida's Farmworkers in the Twenty-first Century, text by Nano Riley and photographs by Davida Johns (2003)

Making Waves: Female Activists in Twentieth-Century Florida, edited by Jack E. Davis and Kari Frederickson (2003; first paperback edition, 2003)

Orange Journalism: Voices from Florida Newspapers, by Julian M. Pleasants (2003)

The Stranahans of Fort Lauderdale: A Pioneer Family of New River, by Harry A. Kersey Jr. (2003; first paperback edition, 2022)

Death in the Everglades: The Murder of Guy Bradley, America's First Martyr to Environmentalism, by Stuart B. McIver (2003; first paperback edition, 2009)

Jacksonville: The Consolidation Story, from Civil Rights to the Jaguars, by James B. Crooks (2004; first paperback edition, 2019)

The Seminole Wars: America's Longest Indian Conflict, by John and Mary Lou Missall (2004; first paperback edition, 2016)

The Mosquito Wars: A History of Mosquito Control in Florida, by Gordon Patterson (2004)

Seasons of Real Florida, by Jeff Klinkenberg (2004; first paperback edition, 2009)

Land of Sunshine, State of Dreams: A Social History of Modern Florida, by Gary R. Mormino (2005; first paperback edition, 2008)

Paradise Lost? The Environmental History of Florida, edited by Jack E. Davis and Raymond Arsenault (2005; first paperback edition, 2005)

Frolicking Bears, Wet Vultures, and Other Oddities: A New York City Journalist in Nineteenth-Century Florida, edited by Jerald T. Milanich (2005)

Waters Less Traveled: Exploring Florida's Big Bend Coast, by Doug Alderson (2005)

Saving South Beach, by M. Barron Stofik (2005; first paperback edition, 2012)

Losing It All to Sprawl: How Progress Ate My Cracker Landscape, by Bill Belleville (2006; first paperback edition, 2010)

Voices of the Apalachicola, compiled and edited by Faith Eidse (2006)

Floridian of His Century: The Courage of Governor LeRoy Collins, by Martin A. Dyckman (2006)

America's Fortress: A History of Fort Jefferson, Dry Tortugas, Florida, by Thomas Reid (2006; first paperback edition, 2022)

Weeki Wachee, City of Mermaids: A History of One of Florida's Oldest Roadside Attractions, by Lu Vickers (2007)

City of Intrigue, Nest of Revolution: A Documentary History of Key West in the Nineteenth Century, by Consuelo E. Stebbins (2007)

The New Deal in South Florida: Design, Policy, and Community Building, 1933–1940, edited by John A. Stuart and John F. Stack Jr. (2008)

The Enduring Seminoles: From Alligator Wrestling to Casino Gaming, Revised and Expanded Edition, by Patsy West (2008; with new preface, 2024)

Pilgrim in the Land of Alligators: More Stories about Real Florida, by Jeff Klinkenberg (2008; first paperback edition, 2011)

A Most Disorderly Court: Scandal and Reform in the Florida Judiciary, by Martin A. Dyckman (2008)

A Journey into Florida Railroad History, by Gregg M. Turner (2008; first paperback edition, 2012)

Sandspurs: Notes from a Coastal Columnist, by Mark Lane (2008)

Paving Paradise: Florida's Vanishing Wetlands and the Failure of No Net Loss, by Craig Pittman and Matthew Waite (2009; first paperback edition, 2010)

Embry-Riddle at War: Aviation Training during World War II, by Stephen G. Craft (2009)

The Columbia Restaurant: Celebrating a Century of History, Culture, and Cuisine, by Andrew T. Huse, with recipes and memories from Richard Gonzmart and the Columbia restaurant family (2009)

Ditch of Dreams: The Cross Florida Barge Canal and the Struggle for Florida's Future, by Steven Noll and David Tegeder (2009; first paperback edition, 2015)

Manatee Insanity: Inside the War over Florida's Most Famous Endangered Species, by Craig Pittman (2010; first paperback edition, 2022)

Frank Lloyd Wright's Florida Southern College, by Dale Allen Gyure (2010)

Sunshine Paradise: A History of Florida Tourism, by Tracy J. Revels (2011; first paperback edition, 2020)

Hidden Seminoles: Julian Dimock's Historic Florida Photographs, by Jerald T. Milanich and Nina J. Root (2011)

Treasures of the Panhandle: A Journey through West Florida, by Brian R. Rucker (2011)

Key West on the Edge: Inventing the Conch Republic, by Robert Kerstein (2012; first paperback edition, 2022)

The Scent of Scandal: Greed, Betrayal, and the World's Most Beautiful Orchid, by Craig Pittman (2012; first paperback edition, 2014)

Backcountry Lawman: True Stories from a Florida Game Warden, by Bob H. Lee (2013; first paperback edition, 2015)

Alligators in B-Flat: Improbable Tales from the Files of Real Florida, by Jeff Klinkenberg (2013; first paperback edition, 2015)

Printed in the USA
CPSIA information can be obtained
at www.ICGtesting.com
CBHW030606210724
11769CB00003B/3